MW00651914
PATRON 69
P-3A ORION
GROUND POUNDER
TACC
P-3C ORION UPDATE
NETHERLANDS
RESEARCH LABORATORY
PATRON
P-140

The Age of Orion
Lockheed P-3

05

The Age of Orion

Lockheed P-3

An Illustrated History

David Reade

Schiffer Military/Aviation History
Atglen, PA

ACKNOWLEDGMENTS

This book was written by the author based upon raw data gathered from Lockheed and US Navy records, attending military conferences as well as countless interviews and data gathering visits with those who operate the P-3 Orion, while writing articles for Lockheed's *ASW Log* and *Airborne Log* magazines. Additional information is based on the Author's own US Navy experience, knowledge and educational background as well as his recent work as a consultant on the aircraft.

The author wishes to thank the following for information, photographs and interest for this book :

At Lockheed (ie; Lockheed Aeronautical Systems Co./ Lockheed California Co. / Lockheed Ontario Co. / Lockheed Advanced Development Co./ Lockheed Martin Aeronautical Systems Co./ Lockheed Martin Federal Systems etc.) ; Bob Harper (editor Airborne Log magazine), Doug Oliver, Jack Schroeder, Wayne Thompson, Jim Cress, Woody Woodford, Bob Moser, Rich Brunson, Dana Pierce, Jim Parolise, Gerry Sink, Denny Lombard, Mike Williams, Ed Schultz, Bill Vincent, B."Bud" Neubauer, Mike McKinney, Vic Ehlers.

US Navy; Capt. Chalker Brown, Capt. J.D. Roberts, Capt. Dave Hull, Capt. Ted Klapka, Capt. Ray Leonard, Capt. Pappy Boyington, Cdr. Glen Woods, Capt. Ray Figueras, Cdr. Vic Ristvedt, Cdr. Phil Winter, Cdr. Dave Porter, Cdr. Ray Recindes, CDR Rod Carlone, Cdr. Barry Lavigne, Cdr. Jeff Kunkel, Cdr. Dan Squires, Cdr. Dave Martin, Cdr. Kevin Ketchmark, Cdr. D. Dow, Cdr. Kurt Unangst, Cdr. Don McLaughlin, Cdr. Steve Smith, LCdr. Bruce Silberman, Cdr. Benton Gray, Lt. John McClean, Lt. Andy Holtz, Lt. James Johnson, Lt. Scott Stucky, Lt. Bill McCrillis, LCdr. Tom Reck.

and ; Rick Burgess, Pat Dooling, John James, Anthony Mazzone, Arlyne Meyers, Brian Story, Dan Fleagle, Jeff Hall, Steve Hogan, Bob Eiswerth, Bob Chase, John Vincent, Terry Taylor, Rudy Imperial, Bob Livingston, Paul Soriano, John Romer, Glen Igl, Bob Rohr, Chuck Bostwick,James Allen, Steve Bancor, Buddy Makin, Don Crumchio, Steve Hill, Tom Pillion and Greg Silvernagel.

Industry ; Rob Chase, John Hardy, Pete Patrick, Rick Mcinnis, Mike McKenny, Hank Steinfeld, Mike Swanton, David Dreiling, Phil Teal, Ken Wiegand, Tom Linquest, Hank Davis, Jeff Rader, Kevin Call, John Brandon, Mike Garvey, Tim Cudia, Larry Sakre, David Strong, Neil Bennett, Patrick Stringer, Mike Turton, Bob Lunning, Larry Kotz, Peggy Lowdes .

International Operators; (Australia) Air Comdr. Laing, Sqd. Ldr. Warren Ludwig, Sqd. Ldr. Graham Lake, Ft.Lt. Andy Macfarlane, Fl.Lt. Philip Bryden, Wing Comdr. Ken Drover, Fl.Lt. Brendan Dalton, Fl.Lt. John Harwood, Warrent Ofc. Joe Hattley, Howard Harvey. (New Zealand) Sdr. Ldr. Kevin Short, Sgt. Graeme Pleasants. (Norway) LCol. Roy Pettersen, 2nd Lt. Sverre Iver Aastrop. (Netherlands) Cdr. Rob Patijn. (Portugal) LCol. Mendes, Col Jose Pereira. (Spain) Cap. Andres Gamboa. (Japan) Cdr. T. Ikemoto, Cdr. K. Tomimatsu. (Thailand) Capt. Wachira Prateepawanit. (Canada) Col. Herb Smale, Capt. Derek Squire, Capt. Chris Mills, Maj. Al Harvey, Capt. John Blakely, Col. Brian McClain, Capt. Kelly Radcliff. (Chile) Capt. Tomas Schlack. (NASA) Keith Kolher, George Postell, Doug Young. (US Customs) Ed. Price, Sandy Paul, Peter Kendig. (NOAA) Ron Philipsborne, Dave Rathbum, Jim McFadden, Butch Moore, Tim Tomastic.

Photographic Support ; Terry Taylor, Baldur Sveinsson, Marco Borst, Marty Isham, Bob Shane, Milo Peltzer, Bruce and Bob Stewart, David Brown, Mahlon Miller, Harry Repine Jr., John Rossino, Frank McClellan, Ken Schmidt and Georges Van Belleghem and P-3 Publications.

Also Special Thanks to; Grace Harlow-Gilligham, Greg and Mary Ann Miller, Frank Pecora, Don Reade, S. Howell, A. Castledine.

Book Design by Ian Robertson.

Library of Congress Catalog Number: 97-80770

Printed in China.
ISBN: 0-7643-0478-X

We are interested in hearing from authors with book ideas on related topics.

Published by Schiffer Publishing Ltd.
4880 Lower Valley Road
Atglen, PA 19310
Phone: (610) 593-1777
FAX: (610) 593-2002
E-mail: Schifferbk@aol.com
Please write for a free catalog.
This book may be purchased from the publisher.
Please include $3.95 postage.
Try your bookstore first.

CONTENTS

DEDICATION

To P-3 Orion operators worldwide and those crews lost in *...The Age Of Orion...*

574

INTRODUCTION

Ever since a German *Unterseeboot* sank three British cruisers in the opening days of World War I the traditional concepts of sea power have been forever changed and established the formation of a new naval conflict that persists today—the art of Anti-Submarine Warfare.

The ASW concept has been a developing one over the past eight decades and has utilized the most current scientific and technological innovations to counter the threat imposed by the submarine. Technology that has included development of advanced weapons and detection sensor as well as the evolution of the aircraft in the art of submarine hunting. In fact, ASW has been seen as a technological race between development of the submarine itself and the means to counter the threat it represents.

Submarines have twice in the last century almost single-handedly conquered their enemies through the employment of unrestricted warfare on merchant traffic and by controlling key shipping lanes of seaborne commerce. During those periods the submarine was considered underhanded and those aboard pirates to be hung in wartime. But each time, ASW capabilities were employed to counter the U-boat menace and turned the tides of events in both WWI and WWII.

Besides such innovations as the Convoy system and armed merchant decoys known as Q-Ships, ASW technology has also taken the form of weapons such as Depth Charges, Mines and electronic sensors including RADAR and SONAR. During the first world war, ASDICS, an active echo-ranging sonar, was introduced to locate submerged submarines. This system was carried over to the next world war with the addition of radar. Radar took the night away from the submarine, the only convenient time to recharge their batteries. But it was a combination of these electronic ASW systems and development of the aircraft that was to revolutionize the concept of ASW.

Although utilized in the first world war, the anti-submarine aircraft came of age in World War II. At the start of WWII, aircraft utilized airborne radar to catch submarines on the surface at night and high-frequency, direction-finding gear to pinpoint the direction of a submarine's communication signals. Later, towards the end of the war, rudimentary active pinging sonobuoys were developed and utilized by patrol aircraft to locate submerged submarines. These patrol aircraft carried the latest in armament such as the depth adjusted bombs and sophisticated homing torpedoes. Despite the advances made in ASW counter-measures, the sub threat didn't go away at the end of WWII. During the war, the Soviet Union built up the largest submarine fleet in the world. The Soviets recognized the true value of the submarine and its influences towards the outcome of any future conflict. They continued to build submarines, borrowing from technological innovations developed by the Germans. The Soviets soon incorporated these improvements, like the snorkel providing efficient recharging of batteries at periscope depth, into their submarines.

By the early 1950s, the newly formed North Atlantic Treaty Organization (NATO) feared the Soviet Union would deploy its subsea force into the world's sea lanes, having effects greater than Germany did during the last war. Later, with the advent of nuclear propulsion and the incorporation of anti-ship and ballistic missiles onboard submarines, new ASW concepts had to be developed which relied heavily on the technology of the day to counter the persistent submarine threat.

The principal focus of these new ASW concepts was in the area of airborne patrol aircraft. With the potential of hundreds of Soviet submarines throughout the world's seas, controlling the sea lanes of international trade with the capability of attacking US aircraft carrier-based battle groups as well as launching ballistic missiles against major US Cities and military installations—requirements for a fast, long-range airborne platform capable of carrying aloft an array of sophisticated sensors and weapons to locate, detect and track (destroy if necessary) these submarines, was needed.

Although there have been a variety of aircraft utilized for ASW by the US Navy since WWII, there has been only one aircraft used since its introduction in 1962—the P-3 Orion!

OPPOSITE: The P-3 Orion in its element, Submarine Hunting.
Lockheed

Chapter 1
Dawning of Orion

After seventeen years of service, the US Navy's prin ciple land-based anti-submarine warfare platform be gan exceeding its reserve growth potential. By 1959, the P-2 NEPTUNE had added more than 40% of its total gross weight. Coupled with the ongoing evolution of new and improved ASW sensors, to keep up with Soviet submarine technology, the Neptune was fast becoming too heavy to take off.

The US Navy began considering requirements for a replacement to the Neptune earlier in 1956 when it attended a NATO armament committee meeting, which convened to discuss a universal replacement aircraft to service participating member nation's ASW forces to battle the ever growing Soviet sub threat. This meeting led to a competition for a new NATO ASW aircraft that subsequently selected a dedicated platform designed and built for ASW, the Brequet 1150 ATLANTIC.

The US Navy was a principal supporter of the Atlantic, while at the same time establishing its own requirements for a new sub-hunter. These requirements led to the issuing of Navy type specification #146 by the Chief of Naval Operations in 1957, for the establishment of a new land-based ASW and Maritime Reconnaissance aircraft to replace P-2 Neptunes and water-based P5M MARLIN flying boats. There were specific requirements within specification #146 pertaining to delivery schedules and cost constraints that dictated the need of developing an existing airframe off-the-shelf for ASW. There were also added requirements for an aircraft with long-range endurance to support ASW operations far out into the vastness of the Pacific. Hence, the Atlantic fell short of the Navy's TS 146 requirements (It's interesting to note that even after the Navy pulled out of NATO's Atlantic program in 1959, it continued to support the project and actually funded approximately 45% of the aircraft's development, more than any other single member partner that eventually acquired the aircraft.).

In May of 1958, the US Navy announced the winner of a design competition established as a result of the CNO issuing Type Specification #146. Lockheed Aeronautical Systems Company of Burbank, California, won the competition with a design based on an advanced developed version of their Electra commercial transport airliner. Several major aircraft manufacturing companies also submitted proposals tailored around existing aircraft, but those designs sacrificed a number of the requirements specified by the Navy's type spec #146. Only Lockheed's Electra proposal fulfilled the requirements with most of the proposed specifications already inherent in the aircraft.

The Lockheed proposal highlighted the Electra's four turbo-prop (jet) engine technology with operational characteristics for long range, high speed (dash) transit at high altitude and low altitude handling qualities as well as fuel economy at low altitude. The proposal also took advantage of the Electra's structural airframe and size which was more than sufficient to house an extensive array of new and sophisticated detection systems. The proposed aircraft also encompassed inherent superlative airport performance that translated into short landing and takeoff capabilities with the ability to utilize existing navy air fields without the need for runway extension.

The Lockheed proposal also brought with it over half a century of ASW experience from producing the first Navy patrol bomber, the PBO-1 HUDSON, through the PV-1 VENTURA and PV-2 HARPOON to the venerable P2V NEPTUNE. Experience that would eventually establish the premier anti-submarine warfare and maritime patrol aircraft of the next half century—the P-3 ORION.

The original L-188 Electra commercial transport program was less than successful, having experienced structural and power plant problems early in its career, including the loss of three commercial airliners. Although the technical difficulties had all been corrected, confidence in the aircraft by the public had waned and few orders were forthcoming from the airlines who by then were opting for the pure-jet airliner. Offering a version of the Electra as the new generation Navy ASW platform provided a program that has since been much more successful than the Electra.

As the program for a new replacement ASW aircraft continued to develop, Lockheed needed to designate a type and name for the new aircraft. Keeping with its production evolution and tradition of naming aircraft after celestial bodies and/or figures of Greek and Roman mythology, Lockheed followed on that which was established with the P-2 Neptune (Neptune being "king of the sea") and christened their new sub-hunting platform the "P-3 Orion," (son of Neptune) "god of the hunt," or simply THE HUNTER.

The P-3 Orion, although somewhat identical to the Electra, is structurally not the same plane. The Orion is not a modified version of the Electra but a totally new aircraft based on a similar design philosophy that has produced an airframe of fail safe construction with the overall strength and long fatigue life in keeping with the requirements of a combat aircraft.

The Orion differs from the Electra in many ways, including the use of large size structural elements, single-piece forgings, integrally stiffened panels and modern manufacturing techniques encompassing weight savings. The wings are made of a fail safe box beam design and utilizes a semi-monocoque pressurized fuselage. The Orion is equipped with Fowler flaps and hydraulically operated tricycle landing gear and hydraulically boosted control surfaces. The overall structure was substantially strengthened compared to that of the Electra and capable of absorbing the constant loads imposed

A US Navy P-3A Orion taking over the vanguard from the retiring P-2 Neptune. ***Lockheed***

A Lockheed L-188 Electra airliner from which the P-3 Orion as based. ***Lockheed***

by low altitude flying. The airframe structure was also resistant to corrosion with protection against the harsh near-sea environment.

Having won the Navy design competition in 1958, Lockheed was awarded an initial research and development contract for the construction and engineering of a mockup of the proposed aircraft. Lockheed quickly modified an L-188 Electra airframe with a dummy magnetic anomaly detector (MAD) tail boom and underbelly fairing to represent a proposed weapons bay. The company flew the mockup aircraft for the first time on 19 August 1958.

The aircraft was flown again in September 1958 for the Navy's mockup inspection board. The success of this initial trial lead to a pre-production contract to Lockheed in October 1958 for the manufacture of long-lead items such as forgings and complicated tools.

Throughout 1959, the proposed ASW mockup aircraft underwent a transformation and was reworked into the Orion aerodynamic prototype designated YP3V-1. This prototype airframe was shortened by seven feet, more accurate to the production configuration. The deleted section provided a significant weight savings and improved the aerodynamic flying qualities of the aircraft. The prototype also incorporated a number of production components including production power plants, armaments, air-conditioning system, airframe structural changes, and mechanical and electronic installations, as well as most of the planned avionics and detection sensors systems.

The completed YP3V-1 prototype aircraft first flew on 25 November 1959 and later through 1960 began a series of Navy board of inspection survey (BIS) trials to evaluate and refine the new Orion. In October of 1960, the Navy awarded Lockheed a contract for seven production Orions that would subsequently be utilized for a long series of BIS test flights. This led to the first full production P3V-1 taking to the air on April 15, 1961. At the same time, the prototype YP3V-1 was

The Orion aerodynamic prototype (mockup) aircraft ; actually a L-188 Electra (the third Electra off the production line) modified with dummy MAD boom and underbelly fairing to represent a proposed weapons bay. *Lockheed*

The first production P3V-1 / P-3A Orion; seen here in the Navy's later maritime patrol gray and white paint scheme. *Lockheed*

A US Navy P-3A in formation with an Australian Royal Australian Air Force (RAAF) P-2 Neptune. ***Lockheed***

finishing a long series of tests and evaluations. Within a year, six of the first seven production P3V-1 Orions would begin the BIS trials conducted by the service and Flight Test Division of the Naval Air Test Center (NATC) at NAS Patuxent River, Maryland. From 15 April to 16 June 1962 five production Orions (#2 through #6) were evaluated in areas of power plants, avionics, and ASW mission equipment performance, as well as fuel, hydraulics, environmental systems and structural service tests. The planes then went on to be used for another series of flight tests to explore the flying qualities and flight envelope of the aircraft. The seventh production aircraft (#7) was assigned to the Naval Weapons Evaluation Facility (NWEF) in Albuquerque, New Mexico, and was used for stores separation and weapons firing testing. In all, over 585 flights were conducted during the BIS trials totaling 2,521 flight hours. At a post-BIS trial conference, the delivery configuration for the P3V-1 Orion was established.

The first production airframe off the production line never flew. The aircraft became a fatigue test article by Lockheed to determine the service life of the Orion airframe. The plane was held in an iron framework of steel, hydraulic gears and jacks to simulate flight stresses by bending the airframe through simulated shear, torsion and compressions of high "G" force turbulence and landing loads. The design structural tests established the aircraft's service life having flown two successful 7,500 hours fatigue cycles, the first 7,500 hour cycle before the next production airframe ever rolled out of the plant.

On 23 July 1962, VP-8 became the first maritime patrol squadron to receive the new P3V-1. Less than a month later, both VP-8 and VP-44 officially took delivery of the remaining flight test P3V-1 Orions at the NAS Pax River, trading in their old P-2 Neptunes.

By September 1962, the new P3V-1 Orion aircraft were re-designated P-3A Orions. (the YP3V-1 prototype was re-designated YP-3A) Within weeks, these same Orions would become fully operational participating in the quarantine of Cuba as the world watched and held its breath as the Cuban Missile Crisis unfolded.

The P-3A Orion was considered a significantly superior ASW weapons system to that of the Neptune when it was introduced in 1962. The Orion was a spacious aircraft with a great interior volume and the electrical power requirements to permit opium utilization of the most advanced electronic detection systems to date with inherent potential growth for future advanced systems.

The P-3 had more interior space for crew stand-up, walk-through convenience than found in the Neptune despite only being a slightly larger aircraft. No more climbing, crawling and ducking to walk through the fuselage like in the Neptune.

The Orion possesses a pressurized fuselage with an air conditioned-controlled environment for all weather comfort to ease crew stress unlike the Neptune. The P-3 was also designed to transit to its assigned operational area faster, locate its submerged target more quickly (in a larger search area) and hold that contact longer than its predecessor. The P-3 had a more advanced airframe with twice the speed, more weapons capability and fuel capacity than the Neptune, as well as a higher ceiling and a greater tactical capability. All without sacrificing any of the aircraft's low altitude maneuverability performance capabilities and short field takeoff qualities inherent to the modern ASW platform.

Chapter 2
Orion the Hunter

There was a new shape in the sky during October 1962. As the Cuban Missile Crisis heated up, the first Navy P-3 Orions were pressed into service supporting the US blockade of Cuba. The new patrol planes must have been quite a sight to the Soviet sailors as the Orions intercepted and shadowed their every move. The Soviet ships were bound for Cuba with prohibited Nuclear missile components stowed in their holds.

The sight of the modern airplane overhead must have generated a flood of reports back to Soviet fleet headquarters, describing those details discernible from quick fly-overs of their ships. Those reports of the Americans fielding a new patrol aircraft at a time of heated tensions must have added to the amount of uncertainty in the minds of those in the Soviet leadership that were planning their next move in response to the blockade.

What did they think of the P-3? What must have been immediately noticeable was a large modern fuselage probably bristling with new electronic equipment to harass their every move. Four large turboprop engines driving the plane,

The very famous shot (taken from a sequence of film broadcasted on television) of a P-3A Orion challenging a Soviet cargo ship at the height of the Cuban Missile Crisis. ***Lockheed***

chasing them down across the sea and staying with them longer. To the Soviets, the Orions were everywhere, staying with them throughout the night and into the dawn, hinting at the new aircraft's long range endurance capabilities.

The Orion must have sent the Soviets an important message during those early days, that times were changing and that the future of seagoing operations was going to be different from here on out—that there was a new era beginning, and it belonged to Orion "The Hunter."

Soviet sub commander's view through the periscope, Surprise! A P-3 Orion fast approching with bomb bay doors open. ***US Navy via J. Turnbull and G. Woods***

P-3 Model "ALPHA"

When the Navy introduced its new ASW patrol plane to the world in July 1962, it was originally designated P3V-1. But a few months later, in September 1962, the military instituted a new tri-service classification/designation policy. Under the new policy, the Orion was re-designated P-3A or the P-3 Alpha.

The P-3A was an aircraft with a strong wide tubular airframe that incorporated somewhat blunt and stubby lift-efficient wings. Its was powered by four Allison T56-A-10W Turboprop engines with Hamilton Standard constant-speed, full-feather props, which utilized a thrust sensitive detector to auto feather the props if the engine failed. These -10w turboprop engines were water-methanol injected engines, providing increased power to each engine by 11% over its derived Allison 501D (T56-A-1) engine used on the Electra. Of the four engines, #2 and # 3 were equipped with 60 kilovolt-ampere (KVA) generators and engine-driven compressors. The generators provided all of the aircraft's electrical load power as well as power to drive all hydraulic functions, flight control boosters and ice control systems, while the compressors maintain cabin pressurization and air conditioning throughout the fuselage. The fourth engine was also equipped with a 60 KVA generator, but no compressor. This generator had a two speed gear box, enabling the generation of 400 cycle power while operating at reduced speed. This helped during pre-flight procedures in areas of low noise abatement restrictions and as a standby power source in-flight.

US Navy P-3A Orion. ***Lockheed***

P-3A Orion takeoff. ***M.A.Cummings - US Navy***

Engine #1 had no add-on components and was usually the one shut down during on-station time in order to conserve fuel. The turboprop engines generated approximately 4,200 equivalent shaft horse power (ESHP) to drive the P-3A up to 400 Knots, yet provide the aircraft the ability to loiter around 200 knots. The turboprop's high-velocity jet exhaust, although not a major means of propulsion, produced approximately 750 pounds of jet thrust (per engine), providing 7% of the maximum rated engine power at takeoff.

The P-3A fuel capacity was approximately 9,200 US gallons or 25,000 pounds carried in one main fuselage fuel cell and five integral wing tanks, giving the aircraft a maximum range of more than 5,000 nautical miles.

The aircraft's initial weapons capability included torpedoes, bombs, conventional and nuclear depth bombs, with provisions added later for mine laying and guided rockets. The Alpha could carry up to 7,200 pounds in its weapons bay and another 16,000 pounds on ten underwing hard-point stations. The aircraft also has interior storage in the aft section for acoustic sonobuoys and marine smoke/flare markers. The sonobouys were launched via a pneumatic launching system that incorporated three pressurized launching tubes and one unpressurized free fall chute through the floor.

The P-3A Orion was of analog technology, equipped with the latest detection electronics and avionics of its day. It was the first Navy patrol aircraft to be equipped with the ASN-42 Inertial Navigation System. The aircraft also utilized existing APN-122 Doppler and APN-70 LORAN navigation systems, as well as standard civilian airways navigation units (navaids) such as the ARN-52 TACAN unit. The communications suite utilized a variety of single channel select (voice) radios, including ARC-94 HF, ARC-51A/52 UHF and ARN-84 VHF units, and the manual operating TT-264/AG TTY teletypewriter system.

Search equipment on the Alpha included dual mounted APS-80 radar, with antenna located in the nose radome and in the aft tail section just behind the MAD Boom. This pro-

Typical Orion weapons load; including torpedoes, mines, missiles and rockets as well as a variety of sonobuoys and sea (smoke / flare) markers. ***Lockheed***

A P-3A taking care of business. ***Lockheed***

The P-3A's tactical compartment. ***Lockheed***

A close-up of the P-3A's radar operators position. ***Lockheed***

vided a full 360 degree radar picture without the need of an externally mounted ventral radome like other patrol aircraft of its day. Other electronics encompassed the ALD-2 electronic counter measures (ECM) direction finding system (used to locate signals in the electromagnetic spectrum, i.e., radar emissions), ASQ-10 MAD detector, the ASR-3 diesel exhaust "sniffer" detector (used to detect the exhaust fumes of snorkeling diesel submarines) and the AQA-3A JEZEBEL long-

A P-3A TACNAVMOD on the attack. ***Lockheed via VPI***

A Navy P-3A firing a Martin ARW-77 Bullpup guided missile. ***US Navy***

range (passive) acoustic signal processor/recorder system and its associated ASA-20 JULIE "explosive" (active) echolocating system.

The ECM system encompassed in-the-wingtip mounted antennas for 360 degree coverage. The AQA-3A Jezebel passive acoustic system, installed in all the early production, flight test aircraft, was later upgraded to the AQA-4A configuration in the remainder of the P-3A production. This Jezebel acoustic system allowed the one acoustic operator to monitor four sonobuoy/sonogram chart displays in limited modes.

The aircraft was established with 12 crew stations, which included the pilot, copilot and flight engineer in the flight station and a tactical crew of five seated at a bench console area equipped with single function displays. This side-by-side configuration ran along the port side of the aircraft amidship. The tactical station bench included (aft looking forward) a Jezebel acoustic operator, the tactical coordinator, or TACCO, the navigator, the Julie/ECM crewman, and lastly, the radar/MAD operator.

A TACNAVMOD P-3A flown by a reserve patrol squadron. ***Lockheed***

A formation of P-3A in route to Vietnam to participate in "Operation Market Time." ***US Navy***

There was a radio operator located behind the cockpit on the portside, who also doubled as the portside observer. The three remaining crew positions were designated observer stations—one opposite the radio operator on the starboard side and one on each side of the aircraft in the back, aft of the main cabin door. Those in the aft observer stations had dual roles as ordnance man (responsible for loading the sonobuoy tubes and ejecting smoke markers and flares) and as in-flight technicians to maintain and repair any of the electronic systems and sensors in flight.

ALPHA MODS

During the course of production, improvements to the P-3A were tested and later either incorporated into production or retrofitted in to earlier aircraft already in the fleet as designed airframe changes.

The first improvement incorporated in to Alpha production was a mine-laying capability. The ASC-5 mine-laying system was incorporated into P-3 150506 and retrofitted into existing production Alphas. The system provided for the aerial mining of enemy waters and harbors to effectively restrict the movement and or destroy an enemie's vessels whether they be surface combatants, merchant vessels or submarines.

The most important improvement to the P-3A was the DELTIC upgrade. The Delayed Time Compression or DELTIC upgrade was a major ASW system update that improved the aircraft's sonar analysis through the installation of the new AQA-5 JEZEBEL passive acoustic signal processor/recorder system. This system improvement provided the single "wet" operator the means to simultaneously monitor and process 16 channels of sonobuoy acoustic data. The upgrade also added an AQH-1 analog acoustic tape recorder to enhance in-flight and post-flight acoustic signal analysis.

There were other minor aircraft improvements included under the DELTIC upgrade program that encompassed enhanced navigation, upgrading the Doppler navigation system to the APN-153A unit and improvements to the ALD-2 ECM direction finding system. The ASA-50 ground speed/tactical bearing computer was also added to the aircraft under the upgrade.

The DELTIC upgrade was also responsible for the deletion of one detection system from the aircraft. The ASR-3 diesel "sniffer" detector was removed, having been considered an obsolete sensor carried over from the Neptune.

Introduced in 1965, the DELTIC improvement program was tested on P-3A 150509 and incorporated into the Alpha production line starting at aircraft 152140, with some earlier P-3A production aircraft retrofitted later.

The next improvement made to P-3 Alphas provided for greater aircraft self-sufficiency through the integration of an Auxiliary Power Unit of APU. The APU generates electrical power for air conditioning and/or engine starting power while on the ground—thus eliminating the need for yellow ground equipment—as well as providing emergency electrical power in flight. Mounted in the forward fuselage, on the starboard side, the incorporation of the APU gave the P-3A additional freedom of operating at minimally equipped bases around the world.

With the engineering design tests conducted on P-3A 152141, the APU Mod was incorporated into the Alpha production line at aircraft 152164. Earlier production Alphas were subsequently retrofitted with the auxiliary power unit and became standard equipment on future Orion models.

Another improvement to the P-3 Alpha model Orion was the addition of the Martin ARW-77 Bullpup guided missile system. This modification gave the P-3A the means to carry and launch Bullpup missiles to use against surface ship or ground targets. Although established into the P-3 Bravo during production, the system was retrofitted back into most P-3A Orions.

Also incorporated into Bravo production and later retrofitted back into all P-3A was the APQ-107 radar altitude warning system, or RAWS. The system worked in conjunction with the aircraft's APN-141(v) radar altimeter and PB-20N autopilot to present the flight deck with a visual and audible low altitude warning.

By the late 1970s P-3 Alphas were serving with reserve VP units. As the capabilities of the computerized P-3C model began proving itself in ASW operations against an ever increasing Soviet submarine threat, the Navy began thinking of a way to modernize its older P-3A Orions comparable with the computerized capabilities of the Charlie Orion. A way to integrate the onboard systems with all new navigation and communications through a general purpose computer was needed to leap from the P-3A's 1960 analog technology to the space-age digitalization of the advanced P-3C of the 1970's. So, in 1975 the Naval Air Test Center (NATC) developed a low cost modification program called Tactical-Navigation modernization program to enhance the P-3 Alpha with digitalization.

The TACNAVMOD program, originally called the P-3A/B update program, followed the P-3C system's design philosophy to tie all the Alpha's avionics together through a computer. The program name was later changed to TACNAVMOD to head off possible confusion with update programs under development for the P-3C.

The program added the AQA-7(v)5 acoustic DIFAR system, including a new sonobuoy receiver set, sonodata processor and tape recorder, as well as expanded the TACCO station adding a new multi-function display with controls for simplified switching between sensor data from the other tactical stations via a ASN-124 data computer. The improvement program also added a new very low frequency inertial navigation system, the ASN-42. This system was similar to the ASN-84 INS on the P-3C but with out its redundant features. Instead the TACNAVMOD navigation system utilizes an OMEGA navigation system as backup. Other improvements include an ASA -66 tactical display in the cockpit.

The TACNAVMOD upgrade, besides linking all onboard systems together, took over the normal routine aircraft computation and book-keeping work of the Alpha. This brought the reserve crews, that had been flying the older Orions, up

A P-3A supporting US Navy patrol boats in Vietnam; Operation Market Time. ***US Navy***

P-3B Orion. ***US Navy***

to the standards of regular active fleet squadrons equipped with the P-3C updates, thus making it easier for reserve crews to locate and track sophisticated Soviet submarines while improving coordination with ground based ASW Operation Centers (ASWOC) and other patrol squadrons during joint operations.

The TACNAVMOD upgrade was retrofitted into P-3A Orions by the Naval Air Repair Facility, Alameda (NARF Alameda—later known as NADEP Alameda) starting in 1979. The first aircraft to be modified was 152157 from reserve patrol squadron VP-64 stationed in NAS Willow Grove, Pennsylvania.

Some TACNAVMOD-equipped Orions were further enhanced in later years. In order to help differentiate between those improvements, designation of a series is in order. The above TACNAVMOD modification, conducted on P-3As and P-3Bs during the late 1970s is designated with the series Block I.

Production of the P-3 Alpha ended in December 1965 with the delivery of 152187 to NAS Moffett Field, California. In all, 157 aircraft and one prototype were produced by Lockheed with many of them still flying. Since their introduction, the P-3A Orion has been constantly modified into various derivative model aircraft tailored for specialized missions and operations. The variant Alphas will each be discussed in Chapter five entitled "Orion the Versatile" and later defined in section A and B of the appendix.

Since its baptism in the middle of the Cold War during the Cuban Missile Crisis, the P-3A continued to be introduced to active fleet VP squadrons. The first Pacific coast unit to receive the new sub hunter was Patrol Squadron forty-six, at NAS Moffett Field on 29 January, 1963. It was VP-46, on 29 January, 1963, that conducted the first west coast deployment of the P-3 Alpha to Adak, Alaska. Later that October, VP-9 made the first West Pacific (West Pac) deployment with the P-3A, operating out of Naha, Okinawa, Japan.

But it was VP-44 that was the first squadron to take the P-3 Alpha abroad, deploying to Newfoundland on 30 April, 1963, although during the Cuban Missile Crisis, VP-8, VP-44 and P-3A operated by Air Development Squadron One (VX-1) flew from short term detachment sites out of Bermuda and the Azores.

The P-3A went on to establish another list of firsts when it was introduced to the Naval Air Reserves in 1970. The Naval Air Reserve began undergoing a reorganization that year, modernizing its fleet of aircraft. It was during 1970 that Naval Reserve VP detachments were established, first in Moffett Field (in July) and then at Pax River (in November). These units were later re-designated and commissioned as reserve patrol squadrons VP-91 (Moffett) and VP-68 (Pax River). Patrol Squadron 91 stood up with the P-3A, having acquired them as a Naval Reserve detachment, while VP-68 transitioned to the Orion from P-2 Neptunes it received when it was a reserve detachment. Subsequently, eleven reserve patrol squadrons would be established over the years and would be eventually equipped with the P-3A.

The P-3 Orion officially went to war in February 1965, supporting US forces in Southeast Asia. Patrol Squadron Nine deployed the first P-3A to participate in the Vietnam Conflict and ushered in an eight-year commitment to anti-infiltration patrols along the 900-mile coastline of South Vietnam in support of Operation "Market Time." These anti-infiltration missions encompassed night radar surveillance to interdict the flow of arms, ammunition and supplies into the south by the Viet Cong. Other P-3 missions in-country included surface surveillance flights monitoring merchant shipping entering North Vietnamese seaports, as well as long range ASW protection of Task Force Battle Groups operating off the Vietnam coast.

Another little known P-3 mission flown during the conflict was that of long-range "Pathfinders." The pathfinding P-3s served as ferry escorts for US Air Force jets making their way to the war in Vietnam. Due to increased aircraft losses in the war, replacement planes were needed as soon as possible. The P-3 pathfinders provided a faster means of deploying replacement jets to the theater. The P-3 would fly with the jets in close formation, providing navigation and communications relay to the transiting aircraft along the way. The P-3 would loiter around while the jets refueled, always ready to assist in case of an emergency.

The first pathfinder mission was flown on January 20, 1966 by P-3A 151381 with subsequent missions flown throughout the conflict including participation from VP-22, VP-4 and VP-9.

The P-3 Market Time operations in Vietnam originated initially out of Naval Station Sangley Point, in the Philippines, then Tan Son Nhut Air Base in South Vietnam and the Naval Air Facility, Cam Ranh Bay, South Vietnam, before finally flying out of NAS Cubi Point in the Philippines. By the war's end in 1973, all active and reserve VP squadrons from both the east and west coasts served in the Southeast Asian conflict, utilizing a mixture of P-3A and P-3B models. The last operational service flights of the P-3A, with a fleet active squadron, took place during November 1978 with VP-44.

The last operational flight with reserve VP units were flown during April 1990. Although the ASW P-3A was withdrawn from fleet service, many derivative Alphas continued to serve in various mission roles, including electronic warfare, weather reconnaissance, airborne oceanographic research and utility transport. These aircraft and their missions will be discussed further in Chapter Five, "Orion The Versatile," and later in the Appendix.

P-3 Model "BRAVO"

Even as the first Orions were beginning to be delivered to VP-8 and VP-44, Navy planners began looking towards the future of the P-3, determining new requirements based on the ever increasing improvements being made in Soviet submarine technology and its effect on crew workload. With new ASW equipment came an increased need for better record keeping and contact information recording during flights, which left little time for concentrating on the tactical requirements of the mission. This led the planners to develop a new system for the P-3 that would interconnect all the various sensor systems on board the aircraft to improve the accuracy of the information collected and therefore permit more time to pursue the tactical portion of the mission at hand. The incorporation of this new integrated ASW system included a number

P-3B Orion. ***US Navy via VPI***

T56-A-14 turboprop engines. ***Lockheed***

of airframe changes to the existing P-3 coming off the production line, which in turn initiated a series of improvements to the P-3A that were significant enough to establish a new model designation. Hence, the P-3B or P-3 Bravo Orion was born.

Right from the beginning, the P-3 Bravo was to have been an interim aircraft and was to have been the design platform for establishing new developments towards a future enhanced aircraft.

Incorporated into P-3 Production, the first model Bravo airframe #152719 (the 158th P-3 Orion off the production line) encompassed Allison's improved turboprop engine the T56-A-14 engine.

The -14 was a more powerful engine with 4,500 ESHP and lacked the problematic water-methanol (alcohol) injection feature. Although rated at 4,500 ESHP, due in part to the lack of structural enhancement to the engine nacelle over the -10, the -14 engine had a considerable excess of power. Tests made with a standard engine nacelle indicated safe operation of the -14 exceeding 5,000 ESHP. This was most useful during takeoffs and landings

The new P-3 Bravo model also encompassed the DELTIC upgrade (and its components), originally introduced into Alpha production, and the APQ-107 radar altitude warning system. The Bravo model also included an auxiliary power unit (APU) originally incorporated into late Alpha production and from here on out a standard feature in the P-3 Orion.

BRAVO MODS

The first improvement to the Bravo model Orion was the (increased) Gross Weight Takeoff modification—later known as the Heavy Weight Mod. This Mod program was initiated, during 1966, to increase the P-3's maximum flight design gross weight or airframe load carrying capability as part of the previously initiated program to improve the ASW systems onboard the Orion aircraft. Under this integrated ASW system program, the decision was made to incorporate redesigned airframe structural elements into production at the earliest opportunity.

This was what led to the creation of the P-3 Bravo model. The increase in weight, some 7,800 pounds (from 126,200 to 135,000 pounds high gross weight), encompassed structural elements throughout the airframe, mostly in the wings to accommodate higher loads of avionics and armaments. The Mod also included beefing up of the landing gear with a "Heavy-Soft" landing gear assembly and the addition of a (non-structural related) emergency fuel-dump capability to the aircraft.

The heavy weight Mod was incorporated into Bravo production with airframe 153442, but three earlier P-3Bs, 152520, 152733 and 152735 were originally modified as design-test aircraft. Although they were equipped with elements of the heavy weight Mod, they were still considered "Light Weight" Bravos (the term used to define those P-3Bs not equipped with the heavy weight Mod) throughout the rest of their service lives. Out of the 144 P-3Bs produced, only the last 63 airframes were structurally strengthened for heavier payloads.

The next improvement to be incorporated into Bravo production was the Bullpup missile system. The Martin ARW-77 Bullpup Guidance (Guided) Missile System gave the P-3B (and subsequent retrofitted P-3 Alphas) an air-to-surface guided missile capability against surface vessels or ground targets. First tested on aircraft 152740, the system was incorporated into production on aircraft 152749 and retrofitted back into all earlier P-3Bs and most P-3As.

The most important improvement modification made to the P-3 Bravo was the TACNAVMOD upgrade in 1970-73. As described before, the TACNAVMOD upgrade digitized the previous analog technology of the P-3 Alpha and Bravo, modernizing them to the capability of the P-3C. The upgrade package added a data processing computer that integrated all onboard systems with a new inertial navigation system with a back-up Omega (worldwide) navigation system and incorporated the advanced AQA-7 (v)5 "DIFAR" capable acoustic system of the P-3C.

Like the Alphas, the Bravos were equipped with the TACNAVMOD as they were transferred from active fleet units into reserve aviation. This Mod was very popular among reserve users and was thought to be tactically a better system package than the P-3C Orions that they were derived from.

In the late 1980s, TACNAVMOD Bravos were further improved. These enhancements included upgrading the acoustic system to the AQA-7(v)7 configuration with new LTN-72 inertial and ARN-99 OMEGA navigation units and the installation of the AAS-36 IRDS infrared detection set, as well as the Harpoon missile launch and control system. This TACNAVMOD upgrade came to be designated TACNAVMOD Block II and was nicknamed "SuperBees." Then, in the early 1990s some Block II TACNAVMOD Bravos were improved again with the incorporation of the new ALR-66(v)2 ESM sys-

P-3B Orions representing different squadrons under Commander Patrol Reserve Wing Atlantic. *US Navy*

tem, SATCOM, new radios and the Bendix APN-234 color weather radar units. These CWR units were later stripped out of the P-3Bs as they were retired and installed on reserve P-3Cs during late 1995 and early 1996. This version of the Bravo came to be known as TACNAVMOD Block III and was nicknamed "KillerBees."

Bravo production commenced during the Navy's fiscal year 1965 (FY65) with the first aircraft delivered to VP-31 fleet replacement (training) squadron on December 13, 1965. Official delivery of the P-3B to active fleet squadrons took place the next month with east coast squadron VP-26 and west coast based VP-9 taking possession of aircraft simultaneously on January 4, 1966.

The Bravo began operational service participating in operation Market Time. It was the Vietnam conflict that would see the Orion's first losses due to combat. During 1968, VP-26 lost two P-3B aircraft and all those aboard them to hostile fire. One was lost in February over the Gulf of Thailand to enemy action, while the second aircraft was shot down in April by fire from a Cambodian gunboat.

The Bravo was also the first of the Orions to be exported to foreign air forces. The first was New Zealand in 1966, then Australia and Norway in 1968. There was a subsequent redistribution of some of the foreign Bravos with Spain and Portugal becoming recipients as well as more recent sales of US Navy surplus Bravos to several nations under foreign military sales (FMS) cases.

P-3 Model "CHARLIE"

From requirements established in 1960 for a new integrated ASW and navigation/avionics system, the Navy began to develop a concept for an improved P-3 Orion early in 1966. The new Orion concept, dubbed the "A-NEW" ASW Avionics System Program, consisted of exploiting the inherent growth capabilities of the P-3 airframe and interconnecting the various ASW systems through a general purpose digital computer. The computerization improved the speed and accuracy of sensor data generation and took over the majority of the aircraft's routine record keeping tasks—freeing up the crew to concentrate on the tactical requirements of the mission.

The program began with a series of initial proof-of-concept flight tests utilizing the YP-3A prototype Orion #148276. The aircraft's ASW, navigation and search sensors were integrated via a centralized computer. The processor was a hand-me-down onboard digital flight computer borrowed from NASA's Apollo manned space program.

P-3B TACNAVMOD Orion. ***Frank McClellan***

P-3C Orion. ***US Navy***

By mid 1966, the Naval Air Development Center (NADC) had designed a flying testbed aircraft, built into Orion #148883, incorporating a new central processor built specifically for the Orion A-NEW program. A UNIVAC CP-901/ASQ-114 central processing system linked the crew and the airborne sensors together to assist in the decision making process. The new computer possessed enough memory to allow for speedy computations and the storage of innumerable amounts of mission details for simplified post flight mission reconstruction analysis. The airborne computer continually updated each sensor operator's display with current mission information, while assuming much of the navigational computations previously done manually, including calculating current geographical and/or tactical position of the aircraft to the target and sonobuoy locations. The processor also assisted the TACCO in directing the aircraft to precise fiy-to-point coordinates for the pilot, as well as the management of current weapons and search stores status. The system even controlled the sonobuoy receiver set and assisted the communications suite to store and transmit message traffic with auto coding/decoding of secure communications in real-time.

The testbed flying laboratory Orion eventually led to the design and production sequence of the A-NEW program and the development of a prototype aircraft that subsequently became a new model of Orion—the P-3C or Charlie.

The P-3C became the first ASW aircraft in history to be equipped with a centralized computer. A P-3B, aircraft #153443 (the 240th Orion airframe), was diverted from the Bravo production line and completed as the prototype YP-3C Orion, incorporating the centralized computer and additional avionics and other changes that define the P-3C model type from the previously produced Orions.

The A-NEW/Charlie Orion featured a more powerful search radar, the APS-115, an advanced acoustic detection/ processing and display system, the AQA-7(v) DIFAR system, and an improved MAD, the ASQ-81(v), as well as an upgraded ECM system (now dubbed electronic support measures, or ESM), the ALQ-78. The aircraft was further enhanced with

YP-3C prototype aircraft. ***Barry Roop via Isham collection***

The first production P-3C Orion. ***Lockheed***

the installation of the AXR-13TV low-light level television system visual sensor (LLLTV) and the addition of secure Link communications via the ACQ-5 Data Link system. The navigation suite was also improved, upgraded with the ASN-84 inertial navigation system (INS) as was most of the aircraft's displays and avionics aboard.

The ASQ-81(v) MAD permitted greater detection of targets at deeper depths. The new MAD is further enhanced by the ASA-65 magnetic compensator, which adjusts the MAD to varying magnetic environmental conditions. It was on November 14, 1975, that a Navy P-3 located the wreck site of the iron ore carrier "Edmund Fitzgerald," lost during a November gale on Lake Superior days before, by MAD.

The APS-115 Radar was a significant improvement over the older APS-80 units in reliability with increased resolution and less interference due to sea return clutter. The APS-115 radar utilized one of the aircraft's new ASA-70 multi-purpose displays located at both the radar operator's (sensor #3) station and at the TACCO station. The ASA-70 displays radar imagery and alpha-numeric tactical data, as well as picture imagery from the LLLTV system.

The AXR-13TV low-light-level TV (LLLTV) system was a visual camera system that could detect surface contacts in low light conditions ranging from dusk to starlit nights or nights with quarter to full moons. The system could be displayed at both the sensor #3 station or at the TACCO station, controlled manually by the sensor #3 operator or automatically by the aircraft's computer.

The new Charlie P-3 had two additional camera systems. One, the KA-74A surveillance camera, was housed in an aluminum container (with four glass windows) attached to the bottom of the forward nose radome. This unit was a gimbal mounted camera with adjustable forward, left or right oblique and verticle (30 degrees down) horizontal views. The other camera system was the KB-18A, mounted on the bottom aft fuselage. It had a 180 degree fore and aft as well as a 40 degree left or right panoramic view, good for assessing bombing results or acquiring intelligence photos of surface targets. Both camera systems could be operated manually by the copilot or automatically by the computer.

The AQA-7(v)1 acoustic processor and display system incorporated the DIFAR (directional low frequency analysis and recording system) passive acoustic signal system. The

P-3C Orion exterior details. ***Lockheed***

Close-up of the P-3C's exterior mounted CAD sonobuoy chutes. *Lockheed*

Close-up of the P-3C's ALQ-78 ESM pod. *Lockheed*

system included for the first time a dual operator station with twin sonogram and video displays. The AQA-7 DIFAR system considerably increased the aircraft's monitoring capability and improved the target bearing calculation ability, resulting in a greater probability of submarine detection. The system also included the AQA-4 multi-track analog acoustic tape recorder with 14 tracks for the recording of acoustic data used in post flight analysis.

The communications suite was improved through the addition of the ACQ-5 Data Link system for the timely transmission and reception of secure Link intelligence and technical data. The system automatically transmits data between relieving aircraft, naval tactical data system ships (NTDS), or tactical support center (TSC) ground stations.

Internally, there were a number of extensive changes due in part to the incorporation of new avionics. These changes included new crew arrangements designed for greater crew efficiency. This comprised a combined navigator/communicator (NAVCOM) station, located on the starboard side aft of the cockpit, across from the old radio operator's station. The new TACCO station was located across from the NAVCOM on the port side, aft of the cockpit—in the old radio operator's station. The sensor three operator was positioned several feet behind the NAVCOM station on the starboard side. The dual acoustic station was located approximately in the same position as on the P-3A/B, port side of the aircraft over the wings. The aircraft retained the two port and starboard aft observer stations.

Other aircraft interior improvements included a ASA-66 tactical display in the cockpit, a new redesigned galley in the aft tail section with a dinette seating area, work bench, dual compartment refrigerator-freezer, a seven meal simultaneous cooking oven and the infamous 36 cup coffee maker. The manually-operated entrance boarding ladder of the P-3A/B was also replaced with an electrically powered folding ladder on the Charlie.

The external differences of the Charlie began with the same basic airframe of the heavy weight Bravo, with the T56-A-14 turboprop engine, but added an improved sonobuoy launching system incorporating 48 externally loaded "A" size, unpressurized sonobuoy chutes—along with 3 "A" size pressurized chutes and 1 "B" size, unpressurized free fall chute installed under the floor in the aft section of the cabin, located aft of the wings. The new sonobuoy system utilized a new, efficient explosive cartridge actuating device, or CAD.

Other noticeable changes included redesigned sonobuoy receiver antennas, the photographic system mounted in a chin pod under the Orion's nose radome, the addition of three windows (one behind the pilot on the port side and two in the galley on the aft starboard side) and the addition of the low-light-level television pod mounted on the starboard side center wing section.

The P-3C model also deleted a number of features from the older P-3A/B Orion aircraft. The Charlie model got rid of the underside mid-fuselage surveillance camera and photo flash cartridge, starboard side wing mounted search light, retro-launcher, SUS-launcher, aft stores observation window (aft section on the starboard side, near the floor) and wingtip mounted ECM antennas replaced by a ALQ-78 ESM pod mounted on the port side center wing section.

In all, 117 P-3Cs were built before the incorporation of the first update. The new P-3C entered service with active fleet patrol squadron 56 in June 1969. The P-3C made its first overseas deployment with VP-49 in July 1970, to NAS Keflavik and the rich hunting grounds of Iceland.

P-3C Update I Orion. *Lockheed*

Patrol Squadron 47 was the first west coast squadron to receive the P-3C, transitioning to the Charlie in October 1970, deploying it for the first time (west coast) to Adak Alaska during June 1971. The first Charlie WestPac deployment took place in August 1975.

P-3C UPDATE I

Ever since its introduction, the P-3C has been continually upgraded through a series of "UPDATES" to keep pace with the technological advances being made by the Soviets in their submarines. The first of these updates began development shortly after the P-3C entered service and never stopped coming.

First tested on P-3C #158928, during April 1974, the "Update I" program added improved computer memory storage to the aircraft through the incorporation of a 4th Logic Unit that included a 393,000 word magnetic drum memory subsystem. Update I also added the installation of the ARN-99(v)-1 OMEGA worldwide navigation system which obtains navigational information via radio signal bursts from a worldwide network of land-based transmitting stations. Other additions included a new ASA-66 tactical display installed between the sensor #1 and sensor #2 acoustic stations (to improve crew effectiveness) and a newer lightweight, more reliable digital magnetic tape transporter.

The Update I program also enhanced several existing systems aboard the P-3C including improvements to the DIFAR acoustic processing system, upgrading it to the AQA-7(V)4 configuration. Other enhancements comprised computer-aided operational software programs to improve the acoustic analysis and signal sorting of the acoustic suite and signal refinement to the ESM system.

The P-3C Update I changes were incorporated into P-3C production during the latter part of 1974 with the first production Update I aircraft #159503 delivered to VX-1 in January 1975.

It should be noted that some of the update I features were later retrofitted into older P-3 Alphas and Bravos as the TACNAVMOD upgrade.

P-3C UPDATE II

Within a few years of the introduction of the Update I into P-3C production, the next improvement program began. This program, established in 1977, was called UPDATE II and became the most comprehensive program to date, enhancing the operational capabilities of the P-3C Orion. Update II added the AAS-36 Infrared Detection Set or IRDS for passive detection of surface objects that emit thermal energy, and the Harpoon missile providing an anti-ship strike capability and a sonobuoy reference system (SRS) for continuous monitoring of sonobuoy positions. The update also added a new acoustic tape recorder and an upgraded acoustic processor, adding the DICASS detection system. The AAS-36 IRDS thermal imaging device, in the category of systems known as forward-looking infrared or FLIR, is a passive thermal scanning sensor that detects high intensity thermal energy emitted by objects against normal background intensities or cooler objects against hotter backgrounds. Being passive and emitting no energy itself, the IRDS system can not be jammed—being that there are no known countermeasures

P-3C Update II Orion. ***Lockheed***

or can not be detected by hostile receivers. The system detects objects in spite of haze, smoke camouflage or darkness. The P-3's IRDS system is installed in a turret housing mounted in the forward nose radome via a retractable mechanism. The system replaces both the KA-74 camera (deleting the nose radome chin housing) and the low-light-level television visual system and its ventrally mounted pod.

The IRDS infrared sensor is useful in passive intelligence gathering and surveillance missions, as well as surface and anti-submarine warfare operations. The thermal imager can locate surface objects (and partially submerged contacts like snorkeling subs), and identify and classify them at standoff ranges. The system has two fields of vision, a wide FOV for searching and a narrow FOV for tracking. The IRDS can be operated manually by the Sensor #3 operator or automatically via auto-track through the aircraft's onboard computer. The IRDS system was further enhanced later with the integration of a video recorder to capture vital thermal imagery, replacing the original film system.

The ARS-3 sonobuoy reference system (SRS) was also added under Update II to aid the ASW system by simultaneously monitoring the positions of up to 32 sonobuoys in the water automatically. The improvement includes a new set of antennas positioned along the aircraft's fuselage.

Update II upgrade also included the incorporation of the AWG-19(v)1 Harpoon aircraft command-launch control system, or HACLCS. This system provides for the launching and control of the AGM-84A Harpoon anti-ship missile, giving the P-3C an all weather, long-range, air-to-surface strike capability.

Close-up of IRDS retractable nose-mounted pod. ***Lockheed***

Other improvements under the Update II program included upgrading the AQA-7 acoustic system with DICASS, the directional, command activated sonobuoy system. DICASS allows for the processing of active pinging sonar signals from DICASS-capable sonobuoys in the water. The data acquired facilitates improved range and bearing of a submerged target. The directional DICASS buoy remains silent until activated by the P-3 acoustic sensor operator via a command activation signal.

Additional Update II improvements encompassed the upgrading the AQH-4 acoustic tape recorder with an enhanced wide-band recording capability that doubles the 14 track capability of the unit it replaces. The new AQH-4(v)2 records, stores and reproduces acoustic data as well as audio signals from the aircraft's ALQ-78 ESM system for post flight analysis.

Begun in 1976, the first production aircraft was delivered to NATC at NAS Pax River in August of 1977. Besides the incorporation of Update II into P-3 production, a number of earlier P-3C non-updated aircraft (known as NUDS) were retrofitted with most of the improvements established under update II. This includes various navigational systems, the IRDS, Harpoon and improvements made to the AQA-7 acoustic processor and display system.

Close-up of AGM-84A Harpoon anti-ship missile. ***Lockheed***

P-3C UPDATE II.5

As the Update II program was being implemented, a number of original avionics systems onboard the P-3 were fast becoming unreliable, obsolete and to some degree unobtainable within the Navy logistics system. As these system replacements became more significant an unscheduled update was coordinated to incorporate all the new improvements into one standardized Mod project. This coordinated upgrade later became known as the UPDATE II.5 upgrade.

Under Update II.5, many of the radio navigational aids were improved and replaced with more reliable systems such as the VIR-31A VOR, ARN-118(v) TACAN and the APN-227 Doppler navigational radar as well as Litton LTN-72 inertial navigation system.

A production P-3C Update II.5 Orion. ***US Navy***

The Update II.5 program also made further improvements to the AQA-7 acoustic processor and display system. The AQA-7(v)10-11 added an interactive control panel for automatic passive tracking. The new update included other additions like the OV-78A integrated acoustic communications system. The IACS provides the aircraft with a capability to communicated with friendly submerged submarines.

Other features of the Update II.5 upgrade included the installation of an OD-159/A auxiliary IRDS display at the TACCO station, a new digital (magnetic) tape set for the loading and extraction of aircraft computer data, the installation of a new compensator system for the MAD and the standardization of wing station stores pylons. The Update also included the incorporation of a new internal communications system (ICS) with loud speakers installed throughout the aircraft. The improvement program also added a fuel venting system for aircraft's fuel cells.

Update II.5 was incorporated into P-3C production with the first aircraft in Navy service by May of 1981.

Plain Jane, non-descript P-3C Update II.5 Orions during a mining exercise (circa. 1986). ***Lockheed***

P-3C Update III Orion. ***Lockheed***

P-3C UPDATE III

Starting in February 1978, the P-3C Update III upgrade program was established and would totally revamp the acoustic capabilities of the P-3C. The upgrade was designed to counter the growing threat posed by the quiet Soviet nuclear submarines.

This new update comprised an entirely new underwater acoustic monitoring system, the IBM "Proteus" UYS-1 advanced signal processor. The UYS-1, replacing the AQA-7(v) 11 DIFAR system, had twice the effectiveness of the Update II.5 Orion with greater sensitivity in detecting underwater submarines. Coupled with the new USQ-78 signal processor, display and control set, which processes, displays and controls data on the acoustic station's two new multipurpose video displays, the UYS-1 doubled the number of sonobuoys that could be monitored. The system also had a built in provision to process future digital array-type sonobuoys as they become available. The advanced acoustic suite also incorporated a new sonobuoy reference system and an advanced sonobuoy communications link receiver. The ARR-78 ASCL set, upgraded the system to simultaneously monitor 32 sonobuoy RF channels.

The new acoustic suite also included an ALQ-158 adaptive control phased array system, designed to reduce the effects of intentional and unintentional interference jamming of sonobuoy RF channels, making it difficult to locate them. The ACPA countermeasures system included the installation of an antenna grouping on the underside of the Orion's fuselage—just aft of the weapons bay.

Other improvements incorporated in the Update III upgrade consist of a new redesigned (transparent) Logic Unit which forms the basis for future data processing modernization.

P-3C Update III Orion. ***US Navy via VPI***

P-3C Update III Orions firing Zuni (unguided) rockets. ***US Navy via VPI***

Actually, due to component obsolescence in the older logic units, the replacement of a new logic sub-system into the onboard computer was established under this update and takes advantage of newer technology to provide for a unit that is smaller, lighter and utilizes less power.

The update also included an improved and redesigned air conditioning system, to help cool the new avionics and an upgraded APU to power the redesigned cooling system. The update III upgrade also included an upgrade of the landing gear to that of a stronger unit.

The production Update III Orion upgrade also incorporates a number of provisions for a second ARR-78 ASCL set to provide the monitoring of an additional 32 sonobuoy RF channels and a second AQH-4(V)2 acoustic tape recorder to accommodate increased sonobuoy monitoring.

Established in February 1978, the Update III upgrade was tested onboard #159889 and later incorporated into P-3C production at aircraft #161762 during May 1984. Delivery of the first aircraft into the fleet occurred during January 1985.

P-3C UPDATE III (Retrofit)

Besides producing production P-3C Update III aircraft, Lockheed was also selected to retrofit many non-updated P-3C (NUDS) and Update I Orions to Update III standards. Lockheed produced modification kits and established field Mod teams to integrate the Update III packages into the candidate aircraft. The first Update III retrofit aircraft was completed in March 1987.

Right from the on-set, the P-3 Orion (in all its models and updated configurations) proved itself as the superior ASW platform it was designed to be to meet the challenges posed by the Soviet submarine fleet.

The aircraft's inherent capability, ease of maintenance and designed growth potential made it a very successful, capable and versatile MPA aircraft. Although mostly devoted to locating and tracking Soviet submarines during the Cold War, P-3 were capable of performing a much wider range of maritime mission tasks. This would become evident in the post Cold War era, but throughout the 1960s, 70s and 80s the P-3 was at the scene of most of the world's international crises.

Besides of its role in the Cuban missile crisis and Market Time coastal patrols during the Vietnam conflict of the 1960s, the P-3 Orion was a major participant in operation Sea Sweep—the search and rescue (SAR) operation to locate fleeing refugee "boat people" escaping the communist domination of South Vietnam.

In 1975, a P-3 in a maritime surveillance role was in direct control of operations during the Mayaguez incident that included one aircraft receiving hits from hostile enemy fire while reporting via radio link directly to the White House. Later P-3s were instrumental in providing fleet support to both the Iranian hostage crisis (1975) and the 1986 raid on Libya.

But the P-3 Orion was foremost a submarine hunter, a mission it performed very well. At the height of the Cold War (1985) the P-3 proved its sub-hunting capabilities successfully during a little known incident that came to be called the "Battle of Bermuda." Due to strong anti-Soviet rhetoric made by both the US and UK at the time, the Soviet Union believed (in the summer of 1985) that it was going to be attacked by

P-3C Update IV Orion prototype. ***Boeing***

the west eventually. The Gorshkuo concept, insistent on winning the battle of the first nuclear salvo, prompted the Soviets to deploy the majority of its surface and sub-surface fleet. From July through September (1985) Navy P-3s flew around the clock operations against this communist armada positioned around the island of Bermuda and along the US coast. At its peak, twenty-eight P-3s operated from NAS Bermuda locating and tracking Soviet submarines. The normal P-3 contingent at Bermuda usually numbered about eight aircraft.

Tensions eased by September, due in part to the vigilance of the P-3. The list of VP participants in the incident including VP-16, VP-24, VP-10, VP-5 VP-45 and even VP-30 fleet replacement (VP training) squadron that utilized the situation for conducting a little on-the-job training.

P-3C UPDATE IV

As the cold war progressed and Soviet submarine technology spawned even quieter submarines, the Navy began to look beyond the capabilities of the P-3C Update III to continue the challenge against the Soviet maritime threat worldwide and that threat imposed by the proliferation of third world diesel submarines.

By the mid-1980s, Navy planners began to establish requirements for a follow-on, next generation anti-submarine warfare aircraft which would include a new integrated avionics package that would come to be called UPDATE IV.

Update IV development began in November 1985 with the Navy's announcement of an open competition for a P-3 upgrade program. The program contract was awarded to the Boeing Company of Seattle, Washington, two years later (July 1987), with a planned completion date scheduled for 1992.

Update IV became the most comprehensive tactical navigational modification to have ever been made to the P-3, even though only one aircraft would subsequently be modified in the end. The upgrade itself was more than a retrofit of sensors and avionics—it included a complete reworking of the tactical compartment, similar to the side-by-side bench configuration of the older P-3A/B Orions.

The Update IV upgrade incorporated a new state-of-the-art acoustic system, the UYS-2 acoustic processor, a new

The tactical compartment of the P-3C Update IV prototype Orion. ***Boeing***

LRAACA / P-7A. ***Lockheed***

imaging radar, the APS-137 inverse synthetic aperture radar, a new digital MAD system and a new passive radar detector—ESM system, the ALR-66(v)5—with improved sensitivity and an interface to provide targeting data to the aircraft's Harpoon missile system. Update IV was originally scheduled to receive the ALR-77 ESM system, but was later replaced with the ALR-66(v)5 ESM system.

The update was also to have included the USC-42(v)3 Mini-DAMA satellite communications system, or SATCOM, as well as new ARC-182 VHF/UHF, ARC-187 UHF and ARC-207 HF Radios. Additionally, the ARN-151 global positioning navigation system (GPS) was selected for the NAVCOM suite. Update IV was also to have incorporated a number of systems carried over from the Update III, including the AAS-36 IRDS, with a new video tape recorder and navigational aids such as dual LTN-72 inertial navigation system, LTN-211 OMEGA, ARN-118 TACAN, APN-227 Doppler units.

At the heart of Update IV was the Boeing's distribution processor/display generator unit, or DP/DGU, based on Motorola MC-68020 processor modules. The system was an integrated data management processing system based on STD-15538 data bus architecture. Instead of one central processor linking all onboard and data storage systems as in pervious updates, each system has its own computer processor linked together via data bus, facilitating faster data processing.

The DP/DGU provided multi-sensor data fusion (accepting data from all sensors) integrating it and displaying it onto new tactical station high-resolution, full-color multi-purpose displays. These universal display and control system (UDACS) displays had multiple window capability, incorporating overlay video, graphics and text tailored to mission needs. The universal displays could be re-configurable to any tactical position to allow the crew to share the workload, thus redistributing tasks to meet changing tactical situations and improve the crew's ability to locate, identify and prosecute underwater targets.

Originally, the Navy plan was to retrofit the Update IV avionics package into approximately 109 Update II and II.5 Orion aircraft. The system was then to have been incorporated into the next generation sub hunter aircraft. The first (and only) Update IV aircraft #160292 was sent to Boeing by the Navy for conversion. Its first flight, as part of a flight test program, was conducted on December 16, 1991. But the program itself was then canceled in early 1992 due to funding reductions related directly to the fall of communism in the east.

LRAACA/P-7A

Shortly after establishing the Update IV program, the Navy initiated another program to eventually replace the P-3C Update III based on requirements for a long-range air anti-submarine warfare capable aircraft. The LRAACA proposed aircraft was to have included the Boeing Update IV avionics package with greater weapons capability, more search stores and greater range, in addition to longer on-station loitering time while using less fuel than the current P-3 Orion. The LRAACA was to have been a quantum leap in ASW technology.

After a design competition for LRAACA, exploring different airframe proposals that included ASW variants of a 757 jetliner, a Gulfstream IV executive jet and the previously mentioned P-3 Orion derivative, the Navy awarded Lockheed a $600 million dollar contract on October 14,1988 to develop and produce the next generation ASW aircraft.

When Lockheed began investigating a future follow-on ASW aircraft for the Navy, it planned on utilizing the newest state-of-the-art avionics and the best equipment of the P-3C Update III to create a superior aircraft for the 21st century.

Lockheed actually proposed a improved version of the P-3C and selected the next available alpha-numeric model designation—the P-3G. But it soon became apparent that an all new aircraft was beginning to take shape and that a new aircraft designation was needed. So, the "G" model Orion was dropped and P-7A designation was adopted for the US Navy's next generation LRAACA aircraft.

The new aircraft that developed, the P-7A, was to have been somewhat larger than the P-3C. It was to have incorporated a 3.5 "G" milspec rated airframe with Update IV avionics, including the APS-137 ISAR radar, ALR-66(v)5 radar detecting ESM system, GPS Navigation, real-time Data Link and SATCOM communications.

The P-7A was also to have included increased sonobuoy capacity (upwards of 112 externally loaded CAD sonobuoy chutes), side-by-side tactical compartments, new technology, fuel-efficient GE-38 turboprop engines with over 5,000 eshp, five fly-by-wire controls and a electronic flight instrumentation system, or EFIS "Glass Cockpit."

Other features included a two man cockpit with computerized warnings, cautions and alerts, a fold-down heads-up display at the pilot's position for launching new aircraft weapons such as the AGM-84A Harpoon anti-ship missile and AIM-9 Sidewinder air-to-air missile as part of the aircraft's new self-defense/counter-measures system. The self-defense package also consisted of a Chaff/Flare/RF Jammer decoys dispensing system, an infrared missile threat warning system, smokeless engine exhaust and engine infrared suppression features, a low-visibility (gray) tactical paint scheme and laser protection screens positioned in cockpit windows. The aircraft was also to have included a commercial color weather radar with passive weather avoidance mode and provisions for two additional tactical stations for future growth.

The P-7A design was to have incorporated new materials, manufacturing techniques and the integration of advanced technologies that included a degree of parts and avionics commonality with the P-3C Update III.

Under the LRAACA contract, upwards of one hundred and twenty-five aircraft were to have been produced for the US Navy beginning in 1992 with the first aircraft delivered by 1994 and would continue into the year 2000. Many foreign countries were interested in co-developing the P-7A in exchange for obtaining production variants of the aircraft. They

P-7A proposed cockpit. ***Lockheed***

included Germany, expecting twelve of the advanced airframes—without update IV avionics—Italy, Japan, Australia and the United Kingdom. But due to development problems at Lockheed and US Navy budget cuts, due in part to the perceived end of the Cold War with the demise of the Soviet Union, the $600 million dollar development contract and potential sales contracts running upwards of $5 billion dollars for one hundred and twenty-five aircraft was canceled in early 1990.

ORION II

After the cancellation of the P-7A, both Lockheed and the US Navy began studying options for another future replacement for the P-3C Update III. After the announcement that Lockheed would be resuming production of the P-3C for the Republic of Korea, the means to produce a new variant Orion became available. The new Orion, next generation aircraft was designed to be an extension of the Korean P-3C Update III baseline aircraft incorporating most of the planned avionics scheduled for the P-7A.

Originally, Lockheed was going to utilize the next available alpha-numeric designation to define the proposed advanced Orion variant. Due to its policy not to re-use model designations of canceled programs (ie: P-3G used prior to P-7A) the next model number on the list was the P-3H! The P-3H model Orion designation was short-lived though. As the program advanced, the new Orion aircraft was re-designated ORION II.

The ORION II program was to have encompassed a structurally strengthened airframe utilizing advanced composite materials and manufacturing techniques. The aircraft's avionics was to have included the Boeing Update IV package with the UYS-2 processor, ALR-66(v)5 ESM, ISAR, IRDS, SATCOM, digital MAD and GPS, INS, OMEGA navigation systems along with more recent P-3C Update III systems innovations such as the OASIS over-the-horizon, airborne sensor information system and current survivability/vulnerability improvement package consisting of the new AAR-47 Infrared missile detector, ALE-47 Chaff/Flare/RF Jammer (decoys) counter-measures system and fuel tank fire suppression foam. ORION II was also to have encompassed new weapons capabilities that included the stand-off land attack missile, or SLAM, the Maverick infrared air-to-surface missile and the AIM-9 Sidewinder air-to-air missile.

The advanced aircraft was to have included new modern technology turboprop engines based on the General Electric's T407 or Allison's GMA2100 with Hamilton Standard or Dowty all-composite six-bladed prop.

Other airframe changes include new flooring with the electrical load center relocated under the floor and the incorporation of an adjustable 3-2 man cockpit. The interior arrangement was also to have been changed, incorporating a cluster grouped tactical compartment positioned in the center of the aircraft over the wings. This is similar to the Canadian CP-140 Aurora configuration that will be explored more in Chapter IV "World Class Orion."

Planned provisions for Orion II included an in-flight refueling receptacle, anti-skid (carbon) brakes and an "EFIS" glass cockpit, as well as proposed fuselage and wing (insert) extensions as originally proposed under the P-3H program.

As the 1990s progressed, the future of maritime patrol aviation became increasingly doubtful, this due in part to the vacuum produced by the demise of the former Soviet Union and the perceived lack of need for submarine hunting aircraft. Both the US Navy and foreign operators began a process to carefully re-evaluate their maritime patrol requirements and establish future ones in an era of dwindling defense budgets and military downsizing.

We'll see in Chapter III "Orion, New World, New Mission" that the post cold war era was not the end of the P-3 Orion, but the dawning of a new age—a new age that would lean on the Orion's inherent flexibility and versatility to meet the new emerging challenges of the 21st century.

Chapter 3
Orion - New World, New Missions

In the fall of 1989, after many months of economic reforms throughout the Soviet Block, the Berlin Wall was knocked down between east and west Germany. Besides the human flood pouring out of the east, democracy began filtering back in from the west to strike the final blow that toppled the Soviet empire and collapsed state-sponsored socialism. Within two years, the old Soviet Union was no more, broken up into a number of smaller individual independent states and ex-Soviet Warsaw Pact republics. The era of Peristroika and Glasnost spelled the end of the Soviet's designs of world domination under a socialist ideology.

The Soviet Union was just unable to economically continue the technological race with the US that had begun in the early 1950s with the development of the "H" Bomb, the race for space and continued development of advanced military aircraft, ships, submarines and weapons systems. The P-3 Orion had become a formidable challenge to some of their weapons systems and were critical in the Navy's response to the Soviet's global deployment of nuclear powered ballistic missile and attack submarines.

But with the disintegration of the Soviet Union and the disbanding of the Warsaw Pact an environment was created that brought into question the continued need for the P-3 Orion. As soon as the Wall came down, the US Congress made plans to cut hundreds of billions of dollars from the military budget. In regards to the P-3, money allocated for the proposed follow-on P-7A and the developing Update IV upgrade program was cut from the Navy's budget. In Congress' estimation, the perceived lack of submarines to hunt meant there was no need for anti-submarine warfare aircraft to challenge them. Hence the P-3 Orion was doomed!

In the wake of the Soviet breakup, years of inter-service rivalries between the US Army, US Air Force, Navy and Marines came to a head with suggestions made by the other services to scrap the US Navy's VP fleet and turn some of the Orions into logistical transport aircraft and others utilized as standoff land attack missile (SLAM) platforms or Navy medium range (land-based) in-flight refueling tankers. Other detractors called into question the combat survivability of the P-3 in an anti-air weapons environment. What didn't help matters was the fact that for over thirty years the US Navy justified its funding to purchase P-3 aircraft and upgrade them through the need to keep pace with advancing Soviet submarine technology. In other words, the US Navy promoted the Orion's ASW capabilities too well. When it came time for Congress to cut the Navy's budget, all they understood was that the P-3 hunted submarines that were no longer a threat. The VP community was to be hit hard under the new budget cuts and the Navy was scrambling to reevaluate its mission requirements in the new world of the post Cold War era.

The disbanding of the Soviet Union presented new challenges to the restructuring US Navy VP fleet. But several situ-

P-3C Update III + Orion ready for the future, equipped for a new world and new missions. ***Lockheed***

Gulf War P-3C. ***US Navy via VPI***

ations happened next to turn the potential demise of the Navy's VP community around. One was the obvious need for the different services to start working together and discover the concept of joint operations. Another was the Gulf War with Iraq. It was the Gulf War that impressively demonstrated the Orion's flexibility and multi-mission capabilities that would later establish requirements to see the P-3 continue well into the 21st century.

GULF WAR

On August 2, 1990 the military might of the middle eastern nation of Iraq stormed over the borders of its rich Arab neighbor, Kuwait. This invasion of Kuwait sparked an international response with the establishment of an economic embargo by the United Nations that included resolutions demanding Iraq withdraw from the occupation of Kuwait or face military consequences.

An Iraqi ship, one of the first to be intercepted by P-3 Orion at the start of the Gulf War. ***US Navy via VPI***

OUTLAW HUNTER preparing for another mission. ***Ken Schmit***

As the UN debated how to respond to Iraq's aggression, military forces from a coalition of nations flooded into the region and poised to prevent Iraq from carrying its aggression over to other borders such as those with Saudi Arabia. Within forty-eight hours of the initial invasion of Kuwait, the first American forces in the Persian Gulf to begin operations that wouldn't be officially designated Desert Shield until 7 August, arrived. Those first American forces were P-3 Orions.

Forward deployed to the VP base on Diego Garcia in the Indian Ocean, the US Navy P-3 Orions moved quickly to an island off the coast of the Arab state of Oman. Based at the King Faisal Naval Airbase on Al Marirah Island, the P-3 conducted surface surveillance flights of the Persian Gulf area in preparation for carrier battle groups already steaming towards the region. Upon the battle groups' arrival and initiation of economic sanctions against Iraq, the P-3s began maritime interdiction missions supporting embargo operations in the Gulf to locate and identify merchant shipping. The P-3 would vector in coalition surface elements to intercept suspected violators of the UN embargo and conduct boardings and inspections. A second VP detachment was established weeks after the first on the Red Sea coast at Jeddah Saudi Arabia to provide battle group convoy protection as the fleet passed through the Suez Canal and into the Red Sea.

Here, in the Gulf War, the P-3 Orion's standard on-board sensors came into play to provide capabilities that were vital to the US Navy battle groups. The Orion's search radar was used in coordination with mine counter-measures operations to detect floating mines and those just below the surface that could prove hazardous to coalition ships. During the Iran-Iraq war in the mid to late 1980s, the extensive use of mines caused severe damage to US Navy combatants during armed escort operations of re-flagged Kuwaiti oil tankers as Iran tried to close the Arabian Gulf to oil tankers during the conflict.

The aircraft's Infrared sensor also proved useful. In one incident, an Orion infrared detecting set was able to detect a ship with painted-out Iraqi markings under freshly painted (bogus) Egyptian markings during a maritime interdiction mission. The Orion's IRDS also provided night anti-terrorist protection surveillance of the battle group guarding against possible attacks by small patrol craft and boats. The Orion's communications suite was also invaluable to the fleet, providing COMS-relay tasks during battle group convoy escort movements through the Suez Canal and the Red Sea.

Some of the Orions operating in the Gulf War were equipped with Texas Instruments APS-137(v)1 ISAR radar. The inverse synthetic aperture radar is a stand-off, two dimensional, real-time imaging radar that not only detects surface contacts but can image the contract and compare its outline for accurate identification. The ISAR uses the Doppler frequency developed by the motion of the ship swaying in the ocean to extract information about the shape of the ship to create a picture of its outline and dimensional structures. The ISAR provides extremely accurate targeting and positioning data that can be used to vector in strike elements. ISAR equipped P-3s provided long range defensive surveillance for the battle group searching for hostile Iraqi patrol boats that operated from coastal waterways and the numerous oil rigs that dotted the northern Arabian Gulf.

As Desert Shield gave way to Desert Storm, ISAR-equipped P-3s imaged patrol boats hiding among Persian Gulf oil rigs and called in strike aircraft to destroy them. In fact, several days before the 7 January 1991 start of the Gulf War air campaign against Iraq, ISAR P-3 conducted coastal surveillance along Iraq and Kuwait as part of the coalition's pre-strike reconnaissance to pinpoint military installations.

On one occasion ISAR P-3s detected a group of Iraqi patrol boats and naval vessels attempting to run from the Kuwaiti seaports of Umm Qasar and Basra to the relative safety of Iranian territorial waters. A specially equipped Orion, called OUTLAW HUNTER, vectored in A-6 and F-18 strike aircraft that attacked the flotilla near Bubiyan Island and destroyed eleven vessels and damaged scores more. Of the one hundred and eight Iraqi vessels destroyed during the conflict, fifty-five were attributed directly to P-3 targeting.

Other P-3s involved in the Gulf War provided battle damage assessment (BDA) of battle group strikes, coalition indi-

Iraqi ship kills scored on the nose gear wheel well door on OUTLAW HUNTER. ***Ken Schmit***

cations and warnings (I&W), command, control and communications (C3I) support to battle groups, as well as oil spill surveillance. This oil spill surveillance mission was in response to the environmental terrorism perpetrated by the Iraqi military which dumped millions of barrels of crude oil into the Persian Gulf. Navy P-3s were tasked with reconnaissance of the spreading oil slick. All totaled, by the end of the conflict, US Navy fleet P-3 Orions conducted approximately three hundred and sixty-nine missions equaling over four thousand flight hours.

The provisional cease-fire that ended hostilities in the Gulf War occurred on February 27, 1991. But in the aftermath of Desert Storm, P-3 Orions continued supporting maritime interdiction surveillance missions enforcing the UN's economic sanctions against Iraq.

The Gulf War was a fortuitous event for the P-3 Orion, in that the conflict highlighted the aircraft's non-ASW, multi-mission capabilities to Congress at a time that the future of Navy VP was in doubt. The P-3 had been the work horse of the fleet. Its long range, dogged endurance and mission flexibility were well suited to the Gulf War's operations. The Orion was vital to the fleet in its ability to make the initial detection, reporting and often controlling the attack of hostile surface combatants. Coupled with its Harpoon stand-off, anti-ship missile strike capability, not exploited during the Persian Gulf conflict, the P-3 had become a potent maritime asset to the battle group in future contingency and limited objective conflicts (CALO) operations.

After the Gulf War, the Navy was quick to promote the P-3 as a multi-mission Maritime Patrol Aircraft (MPA), highlighting its flexibility and foregoing any mention of what would become (for a while) a dirty word—ASW. By 1992-93, a new word was emerging from the hallowed halls of the Pentagon, a word that in itself echoed the nature of the Navy's new outlook in its effort to do more with less. A word that was synonymous with joint operations, and all the services working together for a strong defense. That word was "LITTORAL." The Navy definition of Littoral is the near region of land or more apply a sub-region of coastal areas which extends over the beach inland for several kilometers. Littoral Warfare is part of the Navy's new war fighting doctrine "Forward From The Sea" that replaces the long standing maritime strategy doctrine which had been developed to contain the Soviets during a global conflict.

The new Navy philosophy focuses on peacetime forward presence, crisis response and regional conflict management in future small regional conflicts with joint Navy-Marine forces operating forward from the sea. The Navy-Marine joint force is tasked to maneuver to open a war fighting doorway within the littoral region of a conflict and then act in concert with a much larger combined force (made up of Army and Air force) to provide support for the move inland. This kind of combined operation requires interoperability and provisions for a layered defense of operating forces. For the Navy to operate in the confined and congested waters of the near-land environment calls for command and control, surveillance and battle space dominance (air superiority) capabilities. The P-3 provides those capabilities and is one of the key players in the Littoral Warfare scenario. It's an integral component of these operations that provides not only traditional MPA capabilities but new and emerging capabilities.

FORCE REDUCTIONS

Just as the Gulf War began winding down, the US Congress initiated the first in a series of legislation to cut billions out of the military budget. These cuts included force reductions as well as base closings and realignments.

Despite riding high on its exploits during the Gulf War, the Navy VP community took its hits along with the rest of the Navy. Of the combined twenty-four active fleet patrol squadrons and thirteen reserve VP units in operation at the end of the Gulf conflict (1991), force reductions and budget cuts

A Gulf War weary P-3C Update III +; a clear example of the state of the Navy Orion fleet in the wake of military budget cuts. ***US Navy***

TPS testbed aircraft. ***B.Roop via Isham collection***

would slash the VP community to just twelve and then eight during fiscal years 1996-97.

Congressional military downsizing also closed several VP bases including the main west coast VP base at NAS Moffett Field, California (1994) and those naval reserve air forces at Detroit Michigan (1994), as well as NAS North Weymouth, Massachusetts, in 1996. Squadrons located at NAS Barbers Point, Hawaii, are to be relocated to the Marine Air Station Kaneohe as Barbers Point closes. This leaves the VP hubs of NAS Jacksonville, Florida, and NAS Brunswick, Maine, on the east coast and NAS Whidbey Island, Washington, on the west coast the only major P-3 bases. Additionally, several overseas deployment sites such as the Philippines, Guam, Adak, Alaska, Bermuda and the Azores were closed or scaled back.

Base closings mandated by Congress also encompassed US Navy logistical facilities, forcing the consolidation and closure of naval aviation maintenance depots. In regards to the VP community, the major west coast P-3 aircraft rework facility NADEP Alameda was closed and consolidated all P-3 overhaul work to be performed at NADEP Jax at Nas Jacksonville, Florida.

Additional mandated budget cuts forced the realignment of VP operations. Coupled with squadron disestablishment,

TPS testbed aircraft. ***B.Roop via Isham collection***

NATC (now NAWC Pax River) testing flare dispensers. ***Lockheed***

the opportunity for the Navy to retire its existing fleet of P-3B Orions with the reserves was accomplished and consolidated Orion logistics through the incorporation of an all P-3C fleet and commonality between the active and reserve VP communities.

UPDATE III IMPROVEMENTS

Prior to the Gulf War, a number of improvements were slowly being introduced into the P-3C Update III. The first series of enhancements were established under the Command Survivability Program, conducted by VX-1 air development squadron. This program, an out-growth of studies prompted by the addition of the Harpoon missile into the aircraft, got to trial several self-defense systems that would improve the P-3 Orion's survivability in an ever-increasing hostile anti-air environment. Survivability components included a low visibility, all-gray tactical paint scheme, decoy chaff/flare countermeasures dispensers and a prototype infrared missile detection/ warning system.

The all-gray tactical paint scheme was computer designed to make the aircraft less visible to the human eye and infrared sensors. The TPS paint was radar absorbing and reduced the aircraft's cross-section and infrared signature. The TPS scheme was subsequently incorporated into Lockheed P-3C Update III production and carried over into Update III retro-fit modification. Since the early 1990s, the TPS gray paint scheme has also been introduced into P-3 standard depot level maintenance (SDLM) conducted by the Navy NADEPs.

The program also studied air-to-air countermeasures systems such as anti-radar decoy chaff, used against radar guided missiles and flares to counter heat seeking missiles. Several chaff/flare dispenser systems were tested. Some were wing mounted and others were integrated into the airframe. The P-3 has, in the past, been fitted with flare dispensers before. During the Vietnam conflict, a small number of P-3A/ B Orions were equipped with a flare dispenser mounted in the aft tail section, under the horizontal stabilizers on the port side. This flare dispenser provided a measure of self-defense capability against hostile enemy maritime interceptor aircraft.

Another counter-measures systems established under the survivability program was a infrared missile warning system. The threat warning system detects the heat plumes of in-

P-3C Update III ; note nose and aft mounted AAR-47 IR missile warning system detectors and laser screens in cockpit windows. ***P-3 Publications***

P-3C Update III ; note the lack of ALQ-78 ESM pod indicating the alteration to the ALR-66 ESM system. ***US Navy via VPI***

coming IR guided air-to-air or surface-to-air missiles. The survivability program again trialed several systems. One, the ALQ-157 IR missile warning unit mounted amidships on the port and starboard sides of some west coast P-3 Orions, coupled with wing mounted ALE-37 chaff/flare dispensers, while east coast P-3 aircraft were equipped with the AAR-47 IR missile warning system in coordination with the ALE-39 chaff/flare dispensers that were incorporated into the aft section of the inboard engine nacelles. The AAR-47 has now been established as the standardized infrared missile warning system mounted on the port and starboard side of the Orion's nose radome and both sides of the aft tail radome.

The P-3 Survivability Enhancement program also incorporated self-sealing foam fuel tanks providing security against combat damage from small arms and shrapnel. Finally, the survivability program investigated the potential use of air-to-air defense weapons such as the AIM-9L Sidewinder missile. A number of test firings were performed by the Naval Air Warfare Center at Pax River.

APS-137 RADAR antenna. ***Texas Instruments***

It's interesting to note that this was not actually the first time the P-3 was equipped with sidewinder missiles. A little known trio of P-3s carried the sidewinder missile for self-defense during the early 1960s. It's also been rumored that a P-3 actually shot down a Mig jet during the conflict. It now appears that the situation did occur and that one of these specially equipped P-3s, the only Orions to have been equipped with the missile at the time, shot down the jet! These "Black P-3s" will be discussed later in Chapter Five "Orion The Versatile."

The current NAWC tested sidewinder missile system installed on a fleet P-3C Update III includes a fold-down heads-up display in the command pilot's position.

Another defensive measure installed on some P-3s during the late 1980s, but not included in the survivability program were "Laser Dazzlers" screens for the cockpit and some observer windows. Laser Dazzlers were laser devices originally developed to counter optical surveillance systems and were utilized by both US Navy and the Soviet Navy surface and airborne assets. Unfortunately, Laser Dazzlers could also counter another optical surveillance sensor—the Mk 1 eyeball. These lasers could also cause permanent blindness to those P-3 aircrews and were often employed against Soviet surface vessels.

After an incident where several members of a P-3 crew were blinded, the Navy was prompted to develop a countermeasure to the lasers which subsequently incorporated the

An upgraded OUTLAW HUNTER with an improved OTH-T systems. *US Navy*

installation of metal, copper colored screens into the P-3's cockpit windows.

Despite being considered an inhumane weapon in some quarters, the use of Laser Dazzlers was wide spread. The British were known to have deployed a ship-borne device during the Falklands War that has been attributed to a number of Argentine aircraft crashes. These Laser Dazzlers became such a threat to all concerned that the devices have since been banned by treaty except in the advent of declared war.

With a recent climate of budget cuts and dwindling funds, the previous manor of adding improvements into the Orion in orderly updates has given way to a method of system-by-system trials and proof-of-concept testing known as "Continuous Capability Improvements".

One of the first mission systems to be incorporated into the P-3 in this method was the ALR-66($_{v}$)3 ESM system. The system was introduced into the aircraft, beginning in 1986, as a replacement for the ALQ-78 and as an interim system towards an advanced ESM system selected for the Update IV and P-7A follow-on MPA aircraft.

With installations running parallel to Lockheed's field Mod teams retrofitting older P-3Cs into Update III, the ALR-66 system provided an improved capability to intercept radar signals and determine the contact's bearing, as well as input targeting data collected passively into the aircraft's Harpoon missile system. The new ESM system deleted the existing ALQ-78 ESM pod and incorporated wingtip mounted antennas reminiscent of the P-3A/B configuration. The ALR-66($_{v}$)3 was initially problematic and the manufacturer, General In-

OASIS equipped P-3C; note top fuselage mounted GPS and SATCOM antennas. *P-3 Publications*

CP-2044 Computer. ***Unysis Corp.***

struments, reworked the units aboard the aircraft that were equipped and delivered them back as ALR-66A(v)3. Still plagued with integration of the fine directional-finding (DF) antenna, further modifications included klugging the antenna with the Orion's existing APS-115 radar dish antenna—tapping into the radar antenna on a sharing basis. A recent upgrade to the system increased the sensitivity of the wingtip antennas and re-designated the system to the ALR-66B(v)3 ESM configuration. It is this version of the ESM system that is to be incorporated into the whole Orion fleet and numerous aircraft have already been provisioned (wired) for it.

One of the most important add-on system improvements to be made to the P-3C Update III has been the incorporation of the Texas Instruments APS-137(v)2 Inverse Synthetic Radar. The ISAR radar is a two-dimensional imaging sea-search radar with greater periscope detecting capability. The system utilizes the Doppler effect created by the motion of the object swaying in the ocean to gain information that includes an outline of the bobbing ocean target.

The APS-137(v) radar was an original component of Lockheed's S-3B Upgrade program and versions have also been used on US Coast Guard C-130 Hercules via its counter-narcotics upgrade. The ISAR radar has also been incorporated into the US Navy's HS-60 Seahawk helicopter upgrade.

Introduced into a small number of P-3C Update III Orions prior to the Gulf War, the system became combat proven in the conflict and prompted the Navy to establish a requirement for the whole Orion fleet to be equipped with ISAR. Since the Gulf War, almost half the Orion fleet have been equipped with the new radar system selected as a major component of a new P-3C Update III enhancement program, called ASUW

Adriatic P-3C Orion. ***US Navy***

Improvement Program or AIP, to improve the ASUW capability of the aircraft.

Two other unincorporated trial system upgrades to the P-3C Update III included the Global Positioning System (GPS) and Satellite Communications system, or SATCOM. The GPS system provides for more accurate and precise navigational positioning via triangulation data from geo-synchronous orbiting satellites. The ARN-151 GPS system selected for the P-3 has utilized a variety of antennas mounted on top of the fuselage. They have incorporated a small elevated square shaped fixture resembling a gray elevated bat wing configuration.

The SATCOM system permits clear, uninterrupted voice and data communications transmission worldwide via satellites. The SATCOM antenna has been of a constant design encompassing a white elevated round fixture—looking something like a small toilet seat.

It was just prior to the beginning of the Gulf War that the ISAR radar, GPS and SATCOM components came together to form the basis of a highly specialized P-3 "over-the-horizon" targeting system package installed on a P-3. The prototype aircraft, equipped with the OTH-T system was deployed to the Persian Gulf and was known as "Outlaw Hunter." Developed by the Navy's Space and Naval Warfare Systems Command in cooperation with Tiburon Systems Inc. of San Jose, California, Outlaw Hunter consisted of the APS-137(v)2 ISAR radar integrated with GPS which yielded high-quality targeting data that then could be transmitted immediately (in real-time) to the battle group commander via the SATCOM equipped Officer-In-Tactical-Command Information Exchange System, or OTCIXS. The OTCIXS system permits transmission of the targeting data to the battle-group in order to launch strikes via carrier aircraft or directly to other weapons systems. The OTH-T system can update mission taskings, pass on contact reports, maintain a tactical plot of the battlefield and access battle damage of battle group strikes.

Outlaw Hunter was designed to test the feasibility of the integrated OTH-T system on fleet P-3 Orions. The prototype system evaluated operator workload, engineering problems (i.e. weight, placement of the stand-alone system and antenna locations) and integration with the existing aircraft systems. The aircraft had been undergoing trials in the pacific when Kuwait was suddenly invaded by Iraq. The aircraft was immediately deployed to the conflict and became an integral part of maritime patrol operations conducted against the Iraqi Navy. The highlight of Outlaw Hunter's participation occurred within hours of the start of the coalition air campaign against Iraqi forces. Outlaw Hunter detected a large number of Iraqi patrol boats in the northern Persian Gulf fleeing for the safety of Iranian territorial waters. The OTH-T equipped P-3C vectored in strike aircraft to destroy the fleeing vessels and later provided the BDA of the action, which became the first naval engagement of the conflict.

Since the Outlaw Hunter prototype aircraft, two other P-3C Update III aircraft have been similarly equipped with a follow-on OTH-T system known as OASIS, or Over-The-Horizon Airborne Sensor Information System. The OASIS program was designed to scale down the prototype OTH-T system and incorporate it into the TACCO station. In the meantime, several enhanced versions of the original Outlaw Hunter configuration have been developed and installed into P-3s and have been designated OASIS I and OASIS II. All of these aircraft have been upgraded to the current OASIS III configuration. The proposed future OASIS system, under development, encompasses state-of-the-art technology and software programs to integrate the OTH-T system into the TACCO station. This OASIS system is now a major component of the on-going AIP program and will be integrated into the whole P-3C Update III fleet.

It's interesting to note that the OASIS system has now been proof-of-concept flight tested onboard the S-3B Viking jet (Outlaw Viking) and on the HS-60B Seahawk carrier-borne helicopter (Outlaw Seahawk).

Another important avionics upgrade added to the P-3C Update III was the CP-2044 computer upgrade program which comprised a new digital data processing system to increase power and provided multiple interfaces for an array of new enhanced sensors and display units. The CP-2044 computer upgrade encompasses the installation of the ASQ-212 data processing system to replace the P-3C's older ASQ-114 data processing system and CP-901 computer. Initiated in 1993-94, the processing system has been introduced into the P-3C fleet at the rate of five aircraft per month. The system is a target configuration required prior to installation of the AIP program.

At this time, it must be established that with the inclusion of various avionics improvements mentioned above, the systems configuration of the production/retrofit P-3C Update III has changed. Although substantial changes have been made to the aircraft, no official re-designation of the aircraft has been made by the Navy. This is due in part to the series of individual improvements made to the aircraft and that the new completed configuration is an interim one that is required prior to the initiation of the AIP Improvement Program. For the purpose of clarity within this publication, the unofficial designation of "P-3C Update III+" will be utilized to reflect those Update IIIs that have been improved.

Since the Gulf War, P-3C Update IIIs have been employed in a number of world crises and in the process, discovered new missions and additional capabilities. One of the first was the increased support to the Navy's counter-narcotics operations. The P-3 counter-narcotics mission began in 1990-91 with both fleet and reserve patrol squadrons participating. Conducted throughout the Caribbean, eastern Pacific and Atlantic coastal regions, the P-3 mission includes detecting, localizing and tracking suspected ships and smaller surface craft that fit a particular drug trafficking profile.

The Orion's long-range endurance and day and night surveillance capabilities are well suited to this type of mission. The counter-drug mission incorporates standard MPA

Adriatic P-3C Update III test firing an AGM-65F Maverick air-to-surface missile. ***US Navy***

surveillance procedures to identify suspected vessels sailing in known smuggling transit zones. Although radar (used during the day) and the IRDS (used during the night) are the primary sensors employed, a number of hand held visual surveillance systems have also been used. One, called SIDS-RITS or "secondary imaging dissemination system"—coupled with the "remote imaging transceiver"—is a portable electro-optical system that encompasses a digital camera and laptop transmitter to send stabilized visual data back to operational command centers or surface combatant air controlling units.

As the Orions continued to prove themselves in this role, their tasking was expanded and encompassed, adding a new mission capability to the aircraft. This was accomplished through the installation of the "counter-drug update," or CDU package, which consists of a stand-alone, roll-on/roll-off sensor unit comprised of a APG-66 (F-16) fire-control radar, AVX-1 Cluster Ranger standoff, stabilized high-resolution electro-optical surveillance device and a dual enhanced-communications suite with an interface to the electro-optical system for timely transmission of airborne intelligence imagery. The CDU package provides the P-3C Update III with a new air-to-air intercept capability utilized for the interdiction of drug smuggling aircraft. Initially a small number of Orions were temporarily fitted with a prototype CDU package and rotated through deployments in Panama, Honduras, Puerto Rico and Key West (in Florida) flying patrols in search of drug smuggling aircraft.

With the success of the new air intercept mission capable because of the CDU package, Navy funding was procured to modify more aircraft with the system. A CDU upgrade Mod program was quickly established, consisting of the wiring or provisioning of approximately eighteen P-3C Update III+ Orions and as much as eight production CDU systems produced. Provisioning of aircraft began in 1994 with the production of CDU kits established during the summer of 1995.

One of the more encompassing missions of the P-3C Update III+ since the Gulf War has been the embargo enforcement mission in support of the United Nation sanctions worldwide. Similar to the maritime surveillance flights conducted by the P-3 during Desert Shield/Storm operations against Iraq, the new embargo enforcement missions took on added dimensions that spawned additional mission capabilities and new sensors.

With the break up of the former Soviet Block, many socialist republics have experienced strife as different ethnic or political factions try to gain control of territory. Such has been the case in the former Yugoslavia, where three different republics, Bosnia-Herzegovina, Croatia and Serbia disintegrated into civil war. In an effort to stop the killing inflicted upon the innocent civilian populace, the United Nations enacted a number of resolutions aimed at damming the flow of war fighting materials from entering the region and to restore peace to the war torn republics.

One of those resolutions, #820, imposed an embargo against war materials and arms. The UN tasked NATO to supply maritime assets to assist in the enforcement of the arms embargo, which included the US Navy's participation and the employment of the P-3 Orion.

Both active Navy and reserve patrol squadrons participated in operation "Maritime Guard," which subsequently became operation "Sharp Guard," from the Naval Air Station Sigonella on the Italian Island of Sicily adjacent to the Adriatic sea and the coastal region under embargo. The P-3s' embargo enforcement mission comprised maritime surveillance of the Adriatic and the approaches to the warring republics to detect, identify and query any merchant ships in the region and determine those suspected of carrying prohibited cargos. P-3s were even on the lookout for arms smugglers utilizing cigarette boats and small craft and were to report all suspected vessels to other NATO surface combatant controlling units for potential intercept and boarding for inspection. If found to be violating the imposed embargo, the ships would then be escorted to port for impoundment and disciplinary action.

Most of the these missions included weapons loads usually in the form of MK46 torpedoes and MK20 Rockeye bombs and for the first time operationally used on the P-3, the AGM-65F Maverick missile. The AGM-65F is the infrared guided version of the Maverick air-to-surface missile that provides the Navy with a low cost, anti-ship missile capability for use in the Adriatic. The Harpoon anti-ship missile is an expensive radar-guided weapon and not really suited to the smaller craft employed in the Adriatic. The Maverick missile add-on Mod to the P-3C Update III+ encompasses a control and display setup incorporated into the Sensor #3 operator station, TACCO and co-pilot positions. The Maverick missile was an ideal weapon for the Adriatic in that it was accurate with the ability to see the target, thus reducing collateral damage in its congested waters. Besides maritime surveillance of merchant traffic in the Adriatic, Sharp Guard P-3 monitored the ports along the coasts of Croatia and Bosnia and those of the neighboring republics of Serbia and Montenegro.

Another new mission for the P-3 was further developed during the conflict in the Balkans and included the aircraft going overland. The P-3 Overland Surveillance mission is a strategic surveillance role that provides for immediate intelligence to the battle group commanders and operations commanders on the ground, thus speeding up the battle. The new mission eliminates the subjective intelligence information of previous voice reports based on the interpretation of various onboard (aircraft) sensors and offers more accurate data visually directly to the ground commanders. Like the old adage states, "a picture is worth a thousands words." The overland mission (conducted in the Balkans) consists of monitoring the crisis over the battlefield and relaying intelligence data back to UN commanders in Sarajevo and NATO commanders in Naples, Italy.

During the Yugoslavian conflict, overland P-3s were tasked with monitoring and cataloging various troop movements within Bosnia, locating artillery emplacements and pinpointing other military targets in support of UN/NATO air strikes. The Aircraft also provided battle damage assessment (BDA) after those strikes, as well as enforcing the exclusion zone around Sarajevo. Additionally, coastal patrol flights in the Adriatic were flown to monitor the activities of coastal combatants and peer into local seaports for evidence of embargo violations.

The overland mission utilized both the existing sensors installed on the aircraft (i.e. ISAR Radar and IRDS systems) and the new electro-optical surveillance/reconnaissance system installed into the aircraft. EO systems again are stabilized, airborne visual surveillance devices that provide high-resolution video imagery, detecting and monitoring objects during the day in exceptionally clear conditions—to those in hazy or early dawn periods to late dusk and evenings with quarter to full moon illumination—well outside the range of hostile anti-air defenses. The video imagery can be recorded on tape aboard the aircraft for later analysis or digitally transmitted back to command centers in real-time via line-of-sight tactical link networks or via satellite communications systems.

EO systems have been around since the 1970s when the Pacific Missile Test Center at Point Mugu, California, developed ground-based systems to support various rolling airframe missile tests and close-in weapons trials. The EO systems were further developed and mounted in aircraft to improve air-to-air photographic capabilities to capture imagery of sea-sparrow and sidewinder missile launches as well as Harpoon/Tomahawk missile launches and impact trials. The PMTC EO systems also provided imagery support to US Air Force satellite missile launches at Vandenberg AFB.

The first airborne EO system produced by PMTC was called "Cast Glance II," developed from the ground based Cast Glance device, and was installed onboard a range EP-3A Orion. The system provided a new dimension in airborne optical range surveillance, engineering data acquisition and aerial target test documentation over the existing range photographic systems used to track test objects moving at great speed, which were fast becoming insufficient or ineffective.

PMTC later developed an improved EO system called "Cast Eyes" which caught the interest of the US Navy VP community. EO systems were believed to possibly enhance the fleet MPA P-3C Update III+ Orion's mission capabilities as well as develop new ones. A hybrid version of the PMTC Cast Eyes unit was developed and borrowed by the Navy. It was deployed aboard a fleet P-3C Update III aircraft and trialed in all the Orion's mission profiles. This included the Navy's newest mission, the counter-narcotics surveillance role. This system was nicknamed "Fore cast," a metaphor for the potential use of EO systems in the maritime patrol environment in the future.

The unique capabilities of the hybrid Forecast EO system proved successful in the MPA role and led to greater interest among the Navy's VP community. But ongoing Navy budget cuts at the time hampered development of a dedicated EO system. Only the P-3's counter-narcotics program had authorized funds to develop and acquire new surveillance sensors and presented an opportunity to develop an MPA electro-optical system. The subsequent sharing led to the AVX-1 Cluster Ranger development that is being introduced into the P-3 through the CDU upgrade and the Navy's ASUW Improvement Program, or AIP.

The Cluster Ranger electro-optical system is actually based on another EO system developed by the Naval Air Development Center for the Navy's Special Projects Units in the 1980s and is known as the Tactical Optical Surveillance Systems, or TOSS. The new NADC system was dubbed "Mini-TOSS" and comprises an operator station located in the starboard side aft observer station (the seat turned around facing forward) with the optical sub-system mounted in the TACCO station windows—modified flat to optical quality.

It was after this prototype EO system was produced and installed into the CDU package that the mini-TOSS label was replaced and the new moniker of Cluster Ranger was established. Later, as the conflict heated up in the Balkans, this prototype Cluster Ranger and another Cast Eyes EO system were borrowed and installed on Adriatic deployed P-3C Update III+ and directed overland in Bosnia by the Commander-in-Chief of Naval Operations, Admiral Boorda, to support US/NATO air strikes and "Deny Flight" operations.

Besides providing for the ability of conducting overland and coastal surveillance flights, the advent of electro-optical systems have even made their mark on the more typical Sharp Guard mission. Instead of constantly changing altitudes to "rig" ships—the process by which an aircraft descends to near sea level to note identifying characteristics, name and registry of suspected blockade running vessels—the Orions equipped with EO systems can remain at comfortable altitudes and collect the same information through the long lens of the EO system. There's even data to suggest that eliminating the need to rig ships saves considerable amounts of fuel.

Operation Sharp Guard, which began in 1992, was suspended in December of 1995 when the US brokered a cease-fire and the succession of hostilities. Although P-3 operations out of NAS Sigonella returned to typical post Cold War activities, the EOS equipped Orions continued to conduct overland surveillance flights to monitor the evolution of the Bosnian cease-fire and detect any violations.

All totaled, Navy P-3s along with other international MPA forces amassed 62,300 flying hours, 1,100 consecutive days of continuous maritime coverage with over 54,600 merchant vessels screened and queried. NATO commanders have been quoted as saying that the enforced embargo couldn't have been conducted successfully without the capabilities of the P-3 Orions or crews.

At this point it must be noted that Electro-optical systems are fast becoming a mainstay of future US Navy MPA operations with a production version of the Cluster Ranger, now designated AVX-1, being established as a new component of the Navy's ongoing ASUW Improvement Program to enhance the surface warfare fighting capability of Orion in the future.

The New overland mission of the P-3 is a natural extension of the capabilities inherent in the aircraft. It no way negates any of the Orion's maritime patrol capabilities and with the addition of the electro-optical system onboard actually enhances the existing ASW, ASUW, SAR and coastal patrol surveillance missions.

Another crisis that developed to test the Orion's embargo enforcement mission capability came in the form of operation "Support Democracy," the United Nations imposed economic embargo of the island nation of Haiti. On September 30, 1991, the duly elected president of Haiti, Jean-Bertrand Aristide, was deposed in a coup perpetrated by a military junta. This action initiated a period of unrest within Haiti with diplomatic efforts to return the ousted president proving unsuccessful. This led the UN to pass resolution #841 in June 1993 to enact an economic embargo against Haiti. Initially, its ruling military council announced that they would allow Aristide to return in August of that year, but later reneged and the sanctions were immediately re-imposed during October 1993.

The embargo enforcement operations consisted of a blockade of Haiti by international naval combatants from Britain, France, Canada and the United States, including the US Coast Guard, stopping and searching vessels suspected of embargo violations. Embargo goods included gasoline, oil, oil byproducts, arms and ammunition, police or military equipment, and vehicles and spare parts.

Both active fleet and reserve Navy P-3 patrol squadrons were tasked with maritime surveillance flights of merchant traffic in the area around Haiti. Ships detected were questioned as to their name, nationality, point of origin, destination and contents of their cargos. Any suspicious vessels were then reported, via datalink, to NATO surface combatants acting as aircraft controlling units. The ACUs would then direct available naval or Coast Guard ships to intercept and board suspected vessels for inspection, much like operations in the Adriatic. VP opera-

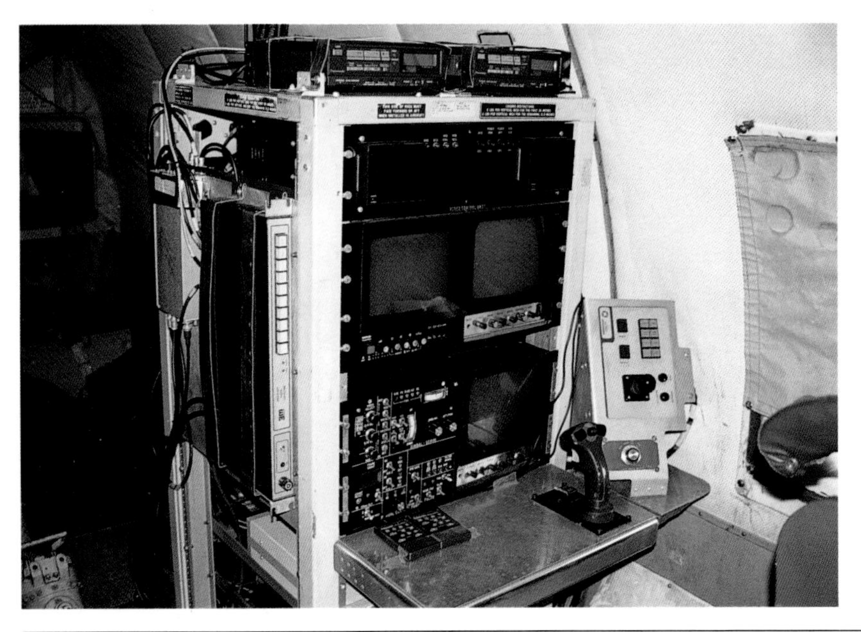

Cluster Ranger's roll-on /roll-off operator station.
P-3 Publications

tions were conducted from NAS Roosevelt Roads, Puerto Rico.

A secondary mission tasking of Support Democracy operations was the location and interception of Haitian refugees fleeing their homeland in upheaval. These operation "Able Manor" flights were flown to the south of the Haitian capitol of Port-Au-Prince and north of the island, in the straits between Cuba and Haiti. The refugees were often found adrift in overloaded, rickety fishing boats or afloat on makeshift rafts. Once rescued, the refugees were subjected to repatriation to Haiti.

Other P-3 operations in Haiti encompassed intelligence gathering missions and overland surveillance flights of Haiti by electro-optical system equipped Orions. As tensions heated up between the military junta and UN/NATO forces, an invasion of Haiti was planned to restore democracy to the island nation. EOS fleet P-3, one aircraft detached from the Adriatic and another CDU aircraft flying counter-narcotics missions out of Panama, conducted pre-invasion intelligence collection. Although there wasn't any subsequent invasion and the crisis was resolved diplomatically, the UN peacekeepers and US troops that eventually came ashore to restore democracy couldn't have done so without the intelligence collected by the overland P-3s. Subsequently, the P-3 went on to support the peacekeeping troops on the ground, covering operations throughout the country.

It must also be noted at this time that the Orion's new embargo enforcement and overland surveillance missions are consistent with the concepts laid out in the Navy's Forward From The Sea doctrine of future joint force in littoral operations. The P-3 will continue to be utilized in these non-traditional roles and is quickly emerging as the primary first on-scene component of the future crisis scenario.

Case in point, in recent low intensity operations throughout the world, the P-3 Orion has been employed principally as the first on-scene asset. In operation "Assured Response," responding to the civil unrest in the West Africa nation of Liberia, Navy P-3 became the vanguard of the operations there. With rival ethnic civil strife endangering US business and diplomatic personnel, the P-3s were first on the scene to provide reconnaissance and surveillance to assess the situation and then supported rescue operations such as guiding special forces helicopters to locate trapped Americans unable to make it to the airport. The Orions also provided critical airspace and communications coordinations, as well as providing EO imagery or a real time picture of the situation there to commanders in control of the operation.

This represents another interesting element of recent P-3 joint operations with other US military services. Once exposed to the capabilities of the Orion, the other branches become dependent on them and request their support there after.

For the purpose of accuracy, it must be noted that although there are a number of P-3C Update III+ aircraft that have received the aforementioned improvements, not all of the Update IIIs have received these add-on systems, nor did the aircraft that were modified receive all the improved systems. It's important to realize, however, that these Update III improvements, i.e. command survivability components (TPS, chaff /flare dispensers, IR missile warning system, self-sealing fuel cell foam, sidewinder missiles) laser dazzler screens, ALR-66 ESM, APS-137 ISAR radar, GPS and SATCOM systems, as well as the OASIS over-the-horizon targeting system with OTCIXS links and the CP-2044 computer, CDU package, new EO systems (Cluster Ranger) and advanced weapons such as the Maverick, created new mission capabilities and enhanced existing ones. Although these systems were introduced into the Orion piecemeal, a P-3 aircraft fleet equipped with all of them would be a formidable force and has led to an all inclusive upgrade to establish these avionics into the Orion. Under an approved requirement to improve the ASUW capability of the P-3, the Navy has established the ASUW Improvement Program.

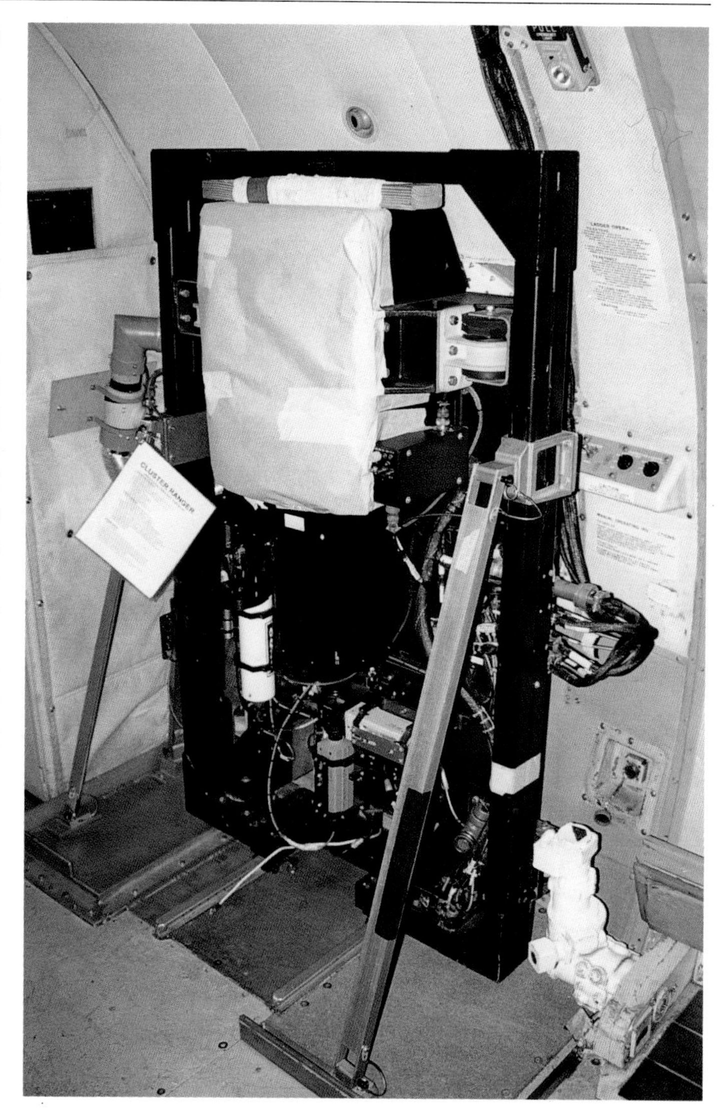

Cluster Ranger's optical sub-system. *P-3 Publications*

AIP PROGRAM

Reminiscent of the P-3C Update projects of the past, the AIP program has been established to enhance the effectiveness of the Navy's large fleet of P-3C Update III configured Orions. This is to be accomplished through the updating of existing systems and onboard sensors and the installation of new system sensors, integrated into the airframe to maximize the plane's ASUW capability and its survivability in the littoral environment of low intensity conflicts.

The focus of the program is to improve the stand-off, over the horizon targeting capacity and interoperability with command, control, communications and intelligence networks and increase the aircraft's ability to receive pertinent real-time tactical data. The program begins with an aircraft target configuration that includes the ASQ-212 data processing group with its component CP-2044 computer (NOW upgraded to the ASQ-222 processor group and CP-2339 computer) and ASN-151 GPS, as well as the ALR-66B(v)3 ESM system or at least wired for the ALR-66 prior to the program initiation.

The avionics improvements scheduled for AIP encompass the replacement of the older APS-115B search radar with a new and improved APS-137(v)5 ISAR multi-mode radar. The APS-137(v)5 version of the ISAR radar encompasses higher resolution and imaging capability to that of ISAR units previously installed on some current P-3s, but adds additional modes including Synthetic Aperture Radar capabilities. The SAR radar has been traditionally an overland imaging radar that uses the motion of the aircraft to generate a Doppler phase history which is processed to form a picture of surface structures and terrain whether it be over land, over water or in some cases over ice during ice reconnaissance. SAR radars are proven overland reconnaissance/surveillance radars and in some cases have air-to-air detection capabilities. The multi-mode APS-137B(v)5 not only provides the P-3 the over water search capability, but an enhanced overland surveillance as well.

AIP also improves the P-3C Update III's ESM system. The new system is an enhanced ESM unit for the passive detection, classification and targeting of electro-magnetic signals generated by unknown contacts. The new ALR-66C(v)3 adds wingtip mounted antennas for 360 degree coverage, as well as a new dedicated spinning fine DF antenna. It deletes the existing ALQ-78 ESM pod antenna of production Update III and does away with the tag-on feedball connection to the radar antenna dish of the previous ALR-66B(v)3 configuration. This is due primarily to the new system's incompatibility to tap-on to the ISAR radar antenna. The new fine DF spinning antenna is to be mounted on the under belly of the Orion's airframe. The small "wart" radome is to be located at the center wing section, forward of the sonobuoy chutes. The new ESM suite is further enhanced with the incorporation of an EP-2060 pulse analyzer, which provides fine frequency analysis and location of contacts from short duration signals.

Another system improvement under the AIP program is to the aircraft's AAR-36 IRDS infrared detection set. The IR unit is to receive an a-focal plane lens modification to increase the existing systems range. This is an interim measure Mod leading to a planned new third generation IR sensor system currently under development. A request for information (RFI) was let by the Navy during June 1996 with the selection of a replacement IR system to be made in early 1997. The new IR system sensor is expected to be incorporated into the AIP program as soon as it becomes available.

Embargo enforcement mission Orions. ***US Navy***

At the heart of AIP is the OASIS III over-the-horizon, airborne sensor information system. This is the third generation of the Outlaw Hunter OTH-T/C^{3}I system that proved successful in the Gulf War with Iraq. The advanced OTH-T system combines ISAR and GPS information to create an accurate tactical plot of the battle area that can transmit vital targeting data on tactical link networks via SATCOM to select tactical command centers. OASIS III can also relay targeting coordinates directly to another strike platform with the appropriate weapons system.

The new OASIS system, now known as a command, control, communications, computer and intelligence, or OTH-T/C^{4}I, is integrated with the AIP's new color, high-resolution universal displays with controls located at the TACCO, NAVCOM and sensor three operator stations. These new universal displays and controls, replacing the P-3s' older monochrome displays, accept and display non-acoustic information (IRDS, ESM and radar) with an interface for the OASIS data.

The aircraft's communications suite is also extensively modified under AIP. The upgrade seeks to minimize the number of components, space, weight and power requirements. The upgrade provides for all tactical links networks (OTCIXS, TADIXS-B, TRAP and dual receive TRE links) to transmit and receive tactical and targeting information without interfering with the aircraft's HF, UHF and VHF radios. The Orion's C^{4}I capability is further enhanced under AIP by the modification of the current wideband SATCOM data and voice system to narrow band SATCOM DAMA unit integrated through a new internal communications system (ICS) that consists of secure voice communications available at more than one tactical station.

Integration of the new and upgraded AIP system is accomplished through the enhanced ASQ-222 central data processing system encompassing the CP-2339 computer. The new CP-2339 computer adds increased processing over the previous CP-2044 model and enormously over the older CP-901 computer existing in the production/retrofit Update III that it replaces. Besides accommodating integration of the planned AIP components, the system encompasses additional interfaces for add-on stand-alone systems (such as EO systems), as well as future development mission avionics.

Under AIP, the AVX-1 Cluster Ranger electro-optical system is introduced into the P-3 as a standardized sensor fit throughout the P-3C Update III fleet. The AVX-1 provides the AIP P-3C with a stabilized, long range stand-off high-resolution optical surveillance imagery that can be transmitted via two different video downlinks, integrated through the central processor and communications suite to ground-based battle group commanders. One is the "Phototelesis" system, an off-the-shelf digital laptop transmission system capable of sending "still" video imagery over standard P-3 Orion link nets.

The other transceiver system is Pioneer. This is a digital video downlink system that transmits color and black & white video imagery in real-time. The ground receiver system is based on commercial computer and video transmission technology integrated with an encrypted safe guard device.

The more new and enhanced capabilities that the P-3C Update III receives, the more formidable it becomes. The more formidable, the more it will be utilized for new and challenging missions creating concerns about the aircraft's survivability in ever increasing hostile environments.

AIP addresses those concerns and standardizes the various survivability countermeasures system on the aircraft that consists of the AAR-47 infrared missile warning system that detects the heat plumes of approaching surface-to-air heat seeking missiles. The system includes detection sensors mounted on both sides of the nose radome and on the aft tail section (there is now a plan to install two additional IR detection sensors on the nose for a total of four). The AAR-47 is further equipped with an auto interface to the ALE-47 chaff/flare/RF jamming decoy dispensing system. The ALE-47 countermeasures dispensing system is utilized in defense of the aircraft, dispersing chaff, flare or RF decoys independently, simultaneously, manually (by the crew), or automatically via the link with the AAR-47. Chaff, small strips of foil, produce multiple radar targets or one large contact to counter radar guided missiles. A chaff cloud can also confuse fire control radars or block missile command signals sent up from ground control stations. Flares produce very hot heat-signatures to counter passive infrared guided missiles. Infrared missiles

AIP Program prototype aircraft ; note ALR-66(V)3 ESM system - DF antenna radome aft of the wings. ***US Navy / NAWC Patuxent River***

will track on the hotter heat source as it drops away from the aircraft. Radio frequency, active jammer decoys or RF jammers interfere with active radar homing missiles by producing a similar noise signature which saturates the missile's receivers.

Installation of the dispensing system includes components fitting into the aft engine nacelles of the aircraft's inboard engines. There has also been consideration of additional dispensers placed in the forward nose section of the aircraft, but there has been no current word as to if it will be done.

Other survivability components of AIP include explosive suppressant foam incorporated into the fuel cells to reduce the aircraft's vulnerability to shrapnel hits or damage from small arms fire and the Maverick missile system integrated into the AIP universal displays and controls, deleting the system's current collection of separate displays installed on some of the P-3C Update III+ Orions.

Another proposed weapon system to be provisioned into the Orion during AIP is the SLAM missile. The Standoff Land Attack Missile provides the P-3 with a precise overland surgical-strike capable weapon used against high-value targets. Designated AGM-84E, the SLAM missile is the land attack version of the internationally popular McDonnell Douglas anti-ship Harpoon missile currently carried aboard the P-3C Update III. The AGM-84E is actually a Harpoon missile body with the imaging infrared targeting head of a Maverick missile. The SLAM missile is also aided by GPS guidance. The missile now provides the P-3 with a stand-off tactical weapon for attacking both land and anti-ship targets while ensuring the aircraft's survivability.

The AGM-84E utilizes the same Harpoon missile launch controls on the P-3C with an additional interface to the AIP's integrated Maverick targeting/controls to assist in the terminal end game. This missile is consistent with and enhances the P-3's overland mission role.

The AIP program is a low cost upgrade comprised of a commercial contract based on non-development, off-the-shelf electronics selected for rapid employment, minimized size, weight and power/cooling requirements. On 19 September 1994, the US Navy announced that the Unisys Corporation-government systems group of St. Paul, Minnesota, was selected as the prime contractor for the AIP program.

The AIP program began during 1996 with delivery of a prototype aircraft to the Navy for testing by summer. This prototype demonstration aircraft was equipped with all or most of the planned upgrade components. Production schedules incorporate seven complete aircraft in 1998 and 63 by the year 2003, and another 146 by the end of the program—although there is hope that eventually the whole fleet would be equipped.

The ASUW Improvement program has obvious benefits beyond the surface warfare capability of the aircraft. The upgrade will provide a new link to other services for closer joint operations and increased interoperability and coordination between surface, sub-surface and other airborne assets without compromising its ASW capability. The AIP program ensures the Navy a truly multi-mission MPA platform far into the future for joint force littoral warfare operations.

MISCELLANEOUS ORION UPGRADES

As the AIP program gets underway, the Navy's P-3 program office is already working on miscellaneous upgrades to the P-3C Update III, some running parallel to the AIP project with others eventually becoming incorporated into the program later. Others encompass improving older update Orions as a means to enhance their capabilities in the interim or prepare the aircraft for eventual upgrading to the Update III similar configuration prior to induction in to the AIP Mod shop.

One of these upgrades includes a CMOS/Bubble extended memory hardware upgrade to make the CP-901 computer in older operational Update II and II.5 Orions work faster. There are also software enhancements to those P-3C Update III aircraft equipped with the CP-2044 computers to provide integration of previously installed stand-alone systems in response to emerging mission requirements.

Other parallel improvements to AIP encompass the installation of new ARC-182 Airways radios and ARC-187 datalink radios as well as an additional upgrade to the P-3C Update III flight station. This project replaces the existing electro-mechanical horizontal situation and flight director altitude indicator gages with new liquid crystal display units. The flat-glass digital LCD display units use existing analog wiring and dash cutouts. The new instruments are being installed into the aircraft during integration of the ARN-151 GPS system which is a pre-requisite of the AIP program.

Another parallel Navy program encompasses replacing the P-3Cs' LTN-72 internal navigation system with the newer LTN-92 INS navigation unit. The LTN-92 INS is an inertial navigation system with an embedded GPS component equipped with reliable laser gyros.

Although the AIP program seeks to improve the ASUW capabilities of the P-3C Update III, the aircraft's acoustic suite is relatively unchanged since the initiation of the production version. In a parallel project, all P-3C NUD, Update I, II, II.5 aircraft acoustic suites are to be upgraded. The ASQ-78A Display and Control Upgrade encompasses modifying the existing ASQ-78 system to be compatible with AIP, since the Update III's UYS-1 acoustic suite is no longer in production. These older NUD and Update aircraft will retain their AQA-7 acoustic systems but will be upgraded with the newer color, high-resolution displays of the AIP Program at the acoustic sensor operator #1 and #2 positions. The upgraded ASQ-78A system fits into the existing ASQ-78 racks and incorporates installation of the new displays. The system has additional interfaces for later integration into the AIP avionics.

Additional proposed parallel improvement projects involve replacing the P-3s magnetic anomaly detector with that of the ASQ-208(v)1 Advanced Digital MAD. The MAD system consists of increased range and digital display output. Originally developed and contracted for the P-7A, the new MAD system has been tested on a NAWC Pax River P-3 and is expected to be installed into future AIP Orions.

One proposed new mission capability for the P-3C Update III AIP configured Orion is the long-range tactical tanker mission encompassing the P-3 in-flight refueling (IFR) tanker modification. The P-3 IFR Tanker Mod capability provides for the refueling of carrier based aircraft far from the CV Battle Group's perimeter. The P-3 IFR Tanker Mod is to be configured as to not degrade any of the Orion's existing capabilities and to have minimal impact to the aircraft. The program consists of modifying the Orion's existing fuel system with new pumps to transfer fuel from the Orion's # 5 fuel tank out to MK32 pods mounted on wing. These pods will house KC-10 type hose-and-drogue basket receptacles. Fuel passes from the P-3's standard fuel cells into #5 tank and out to the wing pods for delivery to thirsty carrier aircraft. The process is designed to maintain the aircraft's balance and stability with minimal impact to the airframe.

There is also consideration for a second capability under the IFR Tanker program, to provide the P-3 with a refueling system to take on fuel in-flight. The P-3 IFR (GIVE) proposal is driving the additional Mod project for an IFR (RECEIVE) capability. This additional capability consists of an in-flight refueling probe to receive fuel from KC-135 Tankers or another IFR equipped tanker.

Several in-flight (RECEIVE) systems had already been flight tested by Pax River during the early 1970s, so the initial engineering and flight test data from those tests were used prior to a new round of evaluations. It's been determined that the program would need minimal funding for prototyping and installation of the system.

Although still under consideration by the Navy, there is currently (as of October 1996) no funding available for the modification program to proceed. The IFR program is a low priority at this time. It's believed that funding could be more forthcoming if concerns that adding the capability to the aircraft won't create an environment where the Orion becomes a very expensive dedicated tanker—much like that experienced by the S-3 Viking community in recent years. As of yet, those concerns have not been addressed.

Looking toward the future and the potential it holds for the Orion. ***US Navy***

One major Navy Orion program that is leading the way for the continuation of the P-3 Orion, encompassing structural and fatigue concerns, is the Structural Data Recording System program. The SDRS is a modification to the P-3 airframe that includes the installation of a series of stain gages and sensors used to profile mission flight dynamics as part of a much larger service life extension program study. The SDRS system records various altitude, weight speed and acceleration parameters that will lead to an understanding of the fatigue characteristics of the Orion. This fatigue data will eventually be applied to load programs that are key to any service life assessment program, or SLAP, that will be established prior to initiation of a SLEP.

In the meantime, as a means to extend the operational service life of the P-3C beyond the next century that is fast approaching, a new Navy program has been initiated to address the high-corrosion areas of the current Orion fleet that have a significant impact on the operational service life of the aircraft.

The program, known as the Sustained Readiness Program or SRP, is designed to replace, upgrade or recondition damaged and corroded airframe components and requires the disassembling of the aircraft in an effort to return the airframes to original or better material condition. The goal of SRP is to restore and improve the Orion's service life and extend the aircraft to thirty-eight years (from the thirty years established by the manufacturer) once the program is complete.

The program includes the reworking of nose gear and landing gears, Wing—internal and external surfaces, flaps, ailerons tabs, fairing and internal center sections, as well as engine nacelles (in and out) APU compartment and the horizontal and vertical stabilizers—practically every component of the airframe.

Besides corrosion areas, SRP addresses aircraft operating systems and components (hydraulics and lift pumps, etc.) that have become obsolete, degraded or in need of replacement. The program also provides for the treatment of predictable corrosion areas and components with new protective coatings to limit future problems.

The SRP contract was awarded to E-systems, Inc. of Dallas, Texas (now a Raytheon company), on 19 September 1994 and runs through the year 2002. The program work is to be performed at the company's Greenville, Texas, division and at the principle sub-contractor's facility, the Northrop-Grumman plant in St. Augustine, Florida.

Information gathered during the dismantling of P-3 airframes during the SRP, coupled with SDRS data collected, will form the basis of a SLAP. The SLAP program will study the Navy Orions to determine their fatigue-life condition and use that data to create preventive measures to extend their service life beyond that established by Lockheed. SRP hopes to generate, possibly another ten to fifteen years of operation. The SLAP recommendations will eventually lead to a SLEP, which will actually implement those recommendations, and extend the aircraft's service life further. The SLEP is expected to begin in the next century.

With a proposed SLEP extending the current Orion fleet past 2017, force level requirements dictate the need for a replacement aircraft for the P-3. If the Cold War hadn't petered out the US Navy would be flying the P-7A or Orion II. But this was not the case, and extending the current fleet of P-3C Update IIIs has become the reality.

Sometime in 2007, the Navy is expected to begin the process to select a replacement MPA program, or MPX. The resulting new next-generation MPA aircraft will be developed and produced, with the first aircraft delivery to be expected by 2015. The MPX is expected to be a commercial competition based on Navy requirements and encompass a variety of aircraft including an new production P-3 variant, advanced commercial aircraft derivative or new development aircraft.

OPPOSITE: An SRP completed P-3C update III that keeps Orion capable far into the next century. ***Photo by Ted Carlson/ Fotodynamics.***

Chapter 4
World Class Orion

The P-3 Orion is not limited to the US Navy, but exists in various configurations throughout the world. Some maritime nations selected the Orion for its superior submarine hunting abilities while others, with limited budgets, chose the P-3 for its inherent multi-mission capabilities.

These nations acquired their new or used (ex-navy surplus) Orions through either foreign military sales via the US Navy or direct from Lockheed's production line—as did the first foreign customer to receive export Orions.

NEW ZEALAND

The Royal New Zealand Air Force became the first foreign Orion operator in 1965 when it placed an order with Lockheed for five DELTIC P-3Bs. Delivered to the RNZAF's #5 Squadron at NAS Moffett Field, California, in the fall of 1966, where aircrews were receiving transition training, the new Bravo Orions provided the squadron the means to manage its vast operational tasks more expeditiously than the Sunderland flying boats that they replaced.

A US Navy P-3C Update III in formation with a Dutch P-3C Update II.5 Orion. ***US Navy***

New Zealand's maritime patrol area covers a broad area of the Pacific. It extends from just north of the Equator down to the South Pole over to the tip of Tasmania, and across to the center of the Pacific. The Kiwi Orion's missions include deployments to Malaysia, Singapore, and as far away as Canada. The self-sufficiency of the Orions is important to #5 squadron, which flies the aircraft mostly from remote island locations and minimally equipped airfields. The isolation of New Zealand's geographical position equates to maritime defense as its number one priority, with ocean surveillance and patrol of its exclusive economic zone (EEZ) a close second on its list of operational responsibilities. ASW training is another important concern to #5 squadron and often includes participating in exercises with other countries such as Australia, Canada and the United Kingdom to gain needed practice due to the lack of ASW experience in its home waters.

Other mission requirements included foreign surveillance flights, providing an aircraft and crews to other Pacific regional island nations. Another important tasking is search and rescue flights throughout the region, which includes emergency medical airlift and support of disaster relief.

The Kiwi Orions also support the RNZAF's maritime strike mission providing standoff targeting as well as ocean navigation and communications relay to RNZAF strike fighters. The RNZAF Orions have also been utilized for civilian research/survey missions such as locating ancient ship wrecks via the aircraft's MAD.

In 1983, the RNZAF initiated an improvement program to enhance the mission effectiveness of their Orions. Called "RIGEL," the avionics modernization program was designed to upgrade the aircrafts' radar, improve the navigation suite, add an infrared detection sensor and incorporate a data handling management/display and control system into the aircraft. The upgrade was an effort to move the Kiwi Orions out of the technology of the 1960s and into the 1990s. Under the program, the aircraft were equipped with an APS-134 radar (a non-imaging ISAR radar) with high-power, digital pulse compression.

The program also included a Litton LTN-72 inertial navigation system, a AAS-36 IRDS infrared detection set and a specialized data handling/universal display and control system. This same UDACS system was at the heart of the Boeing Update IV system and the reason why New Zealand had selected the Boeing company to perform the modification.

Boeing modified one aircraft (#152889) at its Seattle, Washington, production facility between 1983 and 1984 and produced four more Mod kits for domestic installation by Air New Zealand. A sixth P-3B was acquired from Australia in 1985 and it too was later upgraded to the same configuration. With the installation of the new equipment, the RNZAF re-designated their Orion aircraft P-3K. The designation reflects the uniqueness of the aircraft's configuration with an emphasis on their individuality and nationalism. The "K" stands for Kiwi.

In more recent years, the RNZAF has been looking for ways to extend the operational service life of their P-3K, despite a lean economic climate. One such plan was to follow on the heals of RIGEL and add a new acoustic processor and display system, a new ESM system and a GPS unit to the navigation suite while upgrading the data management

A Royal New Zealand Air Force P-3K Orion. ***RNZAF***

Original New Zealand P-3B Orion. ***RNZAF via VPI***

system. The project, called RIGEL II, was to have been started by 1990, but lack of funds canceled it.

The cancellation of RIGEL II has given way to a new proposed upgrade program named SIRIUS. The new project not only encompasses those avionics changes proposed by the RIGEL II program, but also adds a digital MAD system, new HF and UHF radios and an IFF (identify, friend or foe) interrogator. SIRIUS further proposes adding a self-defense system to the aircraft, as well as plans for an EFIS "Glass Cockpit." The program was to have begun in 1996 and run for five years, through the year 2000.

Another project, called "KESTREL," seeks to improve the material condition and fatigue life of the P-3K by re-winging the aircraft. After 30 years of continuous use the Kiwi Orions are beginning to show their age. The wing spar web, upper spar caps, horizontal stabilizers and engine nacelles have begun to experience corrosion and stress corrosion cracks. New Zealand was seriously looking at losing its long range maritime patrol capability with no budget to purchase any new aircraft. Replacing those critical fatigue items on existing aircraft became the only alternative to maintain the MPA mission.

The RNZAF initiated several service life assessment program studies (SLAP) to identify and determine the remaining fatigue life of the aircraft, which included the installation of stress and strain load recorders as well as flight parameters instruments aboard the Orions to generate fatigue data during normal operations. The SLAP data indicated that the P-3K's had less than ten years of service life left and identified which areas needed replacement.

The RNZAF then selected Lockheed to conduct a feasibility study to determine if new production heavy weight P-3C wings, in production for the Korean P-3 program, could be installed on the older (light weight bravo) P-3K Orions. Lockheed's response was a proposed installation plan that consisted of modifying the lower section of the center wing box to that of a P-3C to facilitate the attachment of the new Charlie wings. The upper section of the center wing box remains virtually unchanged. The horizontal stabilizer encompasses a straight forward replacement with the engine nacelles needing only minor refurbishing. The project also included a non-fatigue related modification to the #5 fuel tank to accommodate a new fuel dumping capability.

In the austerity of the late 1990s, project KESTREL is perceived as the most cost effective means for the RNZAF to keep its Orions flying. Starting in 1995-96, the project KESTREL adds another 20 years to the P-3K's service life and provides New Zealand with a maritime patrol asset, vital to the country's defense, well into the year 2015

AUSTRALIA

Australia's Orion experience began in 1968 when it purchased ten P-3Bs from Lockheed. One of the Royal Australian Air Force's two maritime patrol squadrons, # 11 squadron, had

P-3K Orion. ***RNZAF***

Royal Australian Air Force P-3C Orion (#11 Squadron). ***RAAF***

been seeking a replacement for its aging P2V-5F Neptunes. The RAAF was looking for an aircraft that not only provided the current capabilities of the aircraft they were replacing, but also those associated with new emerging mission responsibilities that the current Neptunes couldn't perform.

Although they looked at those maritime patrol aircraft available at the time (the Breguet Atlantic, the not yet available Hawker Siddeley Nimrod and the P-3 Orion), it was the Orion that they chose to replace their Neptunes. The Orion's inherent capabilities convinced them, as did the logistical commonality with the US Navy through the existing Neptune program. The fact that under the military defense program that funded the Orion purchase also included the acquisition of a dozen C-130 Hercules from Lockheed did not hurt either.

The first of ten heavy weight P-3 Bravos were delivered to NAS Moffett field during January 1968. Like New Zealand before them, the Australian aircrews also underwent transition training at VP-31. This transition period was not without incident, however. One of the new Orion Bravos was destroyed on April 11, during acceptance trials before delivery to Australia. P-3B #155296 crash landed at NAS Moffett Field

RAAF P-3C Orion (#10 Squadron) ***RAAF***

The Marconi ASQ-901 acoustic processor and display system onboard Australia's P-3C Orion. ***P-3 Publications***

and burned. Subsequent investigation revealed that the starboard main gear leg detached from the aircraft as it turned onto final approach to land after a series of touch and goes. The gear undercarriage was examined and indicated a fault in the manufacture of the leg forgings were to blame for the gear falling off and causing the resultant crash landing. The destroyed Bravo was later replaced with an ex-US Navy aircraft (#154605).

In 1975, the RAAF initiated another purchase to acquire ten P-3C Orions to replace the last remaining SP-2H Neptunes with #10 squadron (actually, eight aircraft were originally ordered with the sale increased to ten a year later). This marked the first foreign export of the P-3C model, configured to that of the US Navy's Update II configuration.

To keep up with advances being made in Soviet submarine technology, the RAAF initiated a study in 1980 to investigate upgrading their ten P-3B Orions to match the capabilities of their new P-3C Update II. The results of this long study indicated that modifying the older P-3B would not be cost effective, totaling more to modify the older Bravos than it would to buy new aircraft. Eventually, an agreement was reached to purchase new aircraft. This led to the RAAF ordering ten more Charlie Orions in 1982 to replace the Bravos operated with #11 squadron. Under the sales agreement with Lockheed, the original ten Bravos were traded-in to offset the cost of an equal number of advanced P-3Cs. Although the ten aircraft were based on current Lockheed production, configured similar to the US Navy's Update II.5, the RAAF required them to be altered to accommodate the new Marconi ASQ-901 (acous-

RAAF P-3C with toned down markings. ***RAAF***

tic) signal processor/display system and its associated Anglo/Australian Barra sonobuoy system.

These new aircraft were often identified with the P-3W designation; the "W" is used by the RAAF for maintenance purposes to denote slight differences in parts required in repair between the two groups of P-3C aircraft and is not an official type-designation. It should be noted that the original ten P-3C Update IIs were also later equipped with the Marconi acoustic system, but were not maintenance-typed with the P-3W designation.

With a coastline that runs more than 12,000 miles and comprises 2,400,000 square miles of territorial waters—extending from the South Pole (Antarctica) to just north of the Equator, across to the International Dateline to the tip of India down—the RAAF Orions maintain a multi-mission activity. This includes maritime patrol, ocean surveillance, ASW and anti-surface warfare (ASUW) operations, as well as providing support to other military services and government ministries.

The two MPA squadrons also rotate permanent detachments to the Royal Malaysian Air Force Base in Butterworth, Malaysia, providing both aircraft and flight crews to fly Foreign Government "Surveillance Assistance" missions. They also participate in joint military exercises with American forces in the Indian Ocean and South China Sea.

In its civilian responsibility, the RAAF Orions participate in the protection of Australia's Economic Exclusion Zone. The Australian EEZ surveillance flights guard against intrusion by illegal foreign fishing vessels that violate their native fishing zones. One RAAF P-3 is always on quick response (standby) alert for emergency search and rescue (SAR) missions throughout Australia and for customs support when needed.

The end of the Cold War has had far reaching effects on the maritime patrol missions of many nations. The ASW priority is becoming secondary to the traditional MPA roles like ocean surveillance and Anti-Surface Warfare operations. In Australia, where maritime patrol is infinitely vital to its national security, its self-reliance policy gives priority to the military for developing and maintaining capabilities toward the defense of its territories and to promote strategic stability and security throughout the southern Pacific region. As a means of providing continued capabilities toward the defense of its territories and promote stability and security in the southern Pacific, the RAAF has initiated an upgrade program to increase the operational effectiveness and extend the fatigue life of its fleet of nineteen P-3C maritime patrol aircraft.

Proposed AP-3C Upgrade aircraft encompassing exterior modifications associated with the ALR-2001 ESM Mod program. *RAAF*

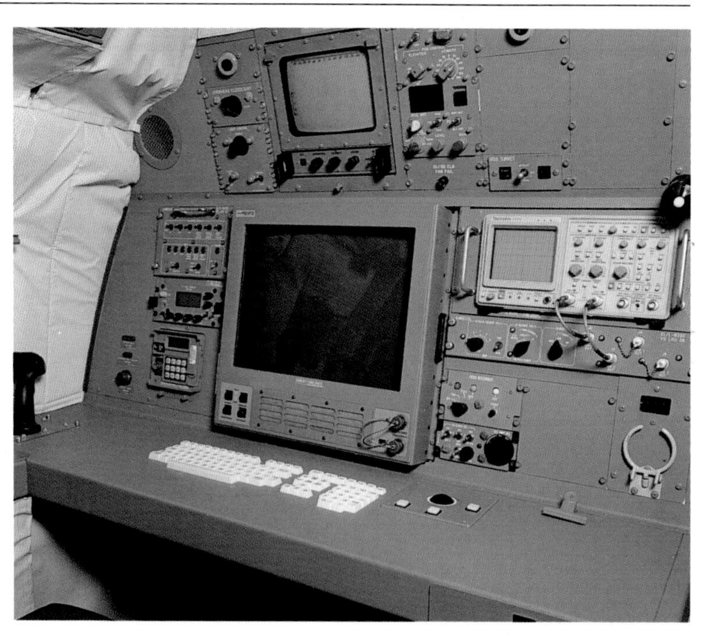

The new Sensor #4 operator station onboard the RAAF P-3C, dedicated solely to ESM management with additional IRDS controls. *RAAF*

After an intense study, the RAAF initiated a aircraft refurbishment program in 1993 to achieve desired improvements to their P-3 Orions. The upgrade program, designated AIR 5276, seeks to optimize the aircraft's radar and infrared sensors while enhancing the acoustic suite. The program also encompasses adding GPS navigation and SATCOM communications capabilities to the aircraft, as well as upgrading the ESM and MAD systems. The program also addresses the fatigue concerns of the aircraft with refurbishment and/or replacement of aircraft structures including older, heavier and less capable, insupportable sensors, avionics and equipment, without sacrificing any of the aircraft's current mission capabilities. The Orion refurbishment program deletes approximately 3,000 to 3,500 lbs of the original RAAF P-3C's payload, which is an important requirement of the program to increase the aircraft's fatigue life.

The new upgrade program's avionics include replacing the existing APS-115B search radar with the EL/M-2022A(v)3 high resolution multi-mode radar system produced by Elta Electronics Industries, LTD. of Israel. The radar is a multi-

Additional underbelly antennas. ***RAAF***

New wingtip mounted receivers. ***RAAF***

mode radar (ISAR, SAR Doppler Beam Sharpening and other target/range profiling modes) capable of periscope detection, target classification with tracking and stand-off targeting capabilities on high and low altitude surveillance profiles. Other features of the radar system include both weather avoidance and navigation assistance modes and are compatible with the onboard tactical data processor, the UNYSIS DDC-060 data management system. The DDC-060 processor is based on an enhanced and expandable version of the CP-2044 computer, which is at the heart of the US Navy's new central processing system, the ASQ-212. The system is a modular unit that includes active windowing capabilities with integrated color, high-resolution sensor (multi-purpose) displays.

The RAAF Orion's existing Marconi acoustic system is to be replaced through the incorporation of the UYS-503 acoustic system produced by Computing Devices of Canada. The UYS-503 is a high speed processing, color discrimination acoustic system, adding a new dimension in signal detection and direction finding. The new acoustics suite is much lighter, constituting the majority of the refurbishment program's weight reduction.

Other program features include the ASQ-504 self-compensating MAD (eliminating the need for high-fatigue stressing MAD-COMP flights), a MAGR 300 (five channel) GPS system and dual H-423 ring-laser gyro inertial navigation sets as well as a communications suite upgrade encompassing new Magnavox (HF, VHF and UHF) radios and SATCOM unit. The communications suite is integrated to provide the operator control and status information through the use of a new control and display system. The upgrade also plans a partial EFIS "Glass Cockpit" Mod on the flight deck.

The avionics upgrade program also incorporates a new ESM system, which was actually introduced into the aircraft beginning in 1992 and was required as part of the target configuration prior to the initiation of AIR 5276. The new ESM system, designated ALR-2001 ODYSSEY and developed by AWA Defense Industries, replaces the existing ALQ-78 ESM system and improves the creation and management of an electro-magnetic tactical surveillance plot. It has the capability for detecting surface ship, submarine, aircraft and land-based transmissions and radar pulses while scanning for hostile weapon targeting systems. The ESM system also assists in passive over-the-horizon targeting for the aircraft's existing Harpoon anti-ship missile system.

The ALR-2001 ODYSSEY component consists of numerous under-fuselage and wing-tip mounted sensor antenna arrays and new interior electronic racks for avionics. The modification is of significant interest as a result of the institu-

Aft tail radome mounted antenna pod. ***RAAF***

tion of a new operator station, solely dedicated to ESM management. The new sensor #4 station is located across from sensor #3 station on a P-3C and includes a 19" telegraphics ESM display and control unit with the controls and display for IRDS. This frees the sensor #3 operator to concentrate on radar and MAD management.

The dedicated ESM operator is a growing trend in maritime patrol aviation. Both the Portuguese Orions and Canadian Auroras have had a two-man sensor #3 station with one operator responsible solely for ESM for years now. The US Navy and other Orion operators are also considering similar ideas for a devoted ESM operator in future enhancement programs. The ESM modification also includes a threat warning display in the cockpit.

Other components of the Mod include numerous antenna arrays housed in the forward radome, aft tail section and wing tip mounted pods and on the underside of the aircraft. The Mod eliminates the existing ALQ-78 ESM pod. Under the ESM program, eighteen of the RAAF P-3C are to be modified starting in 1997.

Although the ESM system Mod project incorporated a new dedicated ESM operator station in the RAAF P-3C, the new upgrade/refurbishment program consists of reorganizing the interior of the aircraft. This includes establishing a new tactical compartment configuration with preliminary plans calling for a cluster grouping in the center of the aircraft, over the wings. This cluster configuration is expected to be a modified version of the Canadian's CP-140 Aurora's "U" shape tactical compartment.

The RAAF upgrade/refurbishment program contract was awarded to E-Systems Inc. of Dallas, Texas, in 1995, with several Australian companies (Aero Space Technologies of Australia, AWA Defense Industries of Australia and Honeywell of Australia) participating in sub-contracting activities. The program is expected to begin in 1997 and run through the turn of the century. Once completed, each enhanced RAAF Orion is expected to be re-designated "AP-3C." The "A" is for Australia!

The RAAF study that indicated the need for an upgrade of RAAF Orions, also indicated that a great percentage of the operational hours of the RAAF P-3C were devoted to aircrew training, logistical support and utility transport. Hence, as an additional effort to extend fatigue life of the RAAF Orions in the future, another P-3 program was developed which resulted in the purchase of three ex-US Navy P-3B Orions.

Under this conversion/modification program, the three aircraft were to be re-configured as TAP-3 Orions. The training/logistics Orion program began with the stripping out of all aircraft acoustic, non-acoustic sensors, avionics, armament, equipment racks and sonobuoy launching/storage system components and associated plumbing from the ex-Navy Bravos.

After standard depot level maintenance and airframe rework was performed, the aircraft were modified with a P-3C Update II.5 flight station instrument configuration to optimize efficiency as a training aircraft . The aft section of the aircraft was modified with reinforced floors to facilitate the addition of floor tracks. The floor tracks run from the new navigators'/aircrew work station (the old radio operators' position) all the way back to the main cabin door. The tracks can handle upwards of 3,450 lbs of cargo or 26 airline passenger seats. Another 1,500 lbs of logistical cargo can be carried in a weapons bay mounted High Capacity Cargo Pannier. The MOD also included new interior wall coverings and trim throughout the aircraft.

Each of the three aircraft are also to be equipped with the APN-234 color weather radar. This commercial CWR unit includes a cockpit display and an antenna mounted into the nose radome.

Results of a recent paint scheme test used to evaluate the best color scheme for the AP-3C program. ***RAAF***

The primary mission of the TAP-3 aircraft is pilot and flight engineer aircraft conversion training, with cargo and passenger transport support to deployed aircraft and flight crews secondary. This decreases the wear and tear on the P-3C Orions, thus increasing their fatigue life.

The three ex-US Navy P-3B were inducted into NADEP Jacksonville in the summer of 1995 with roll out and delivery of the first TAP-3 Orions during the fall of 1996. They will be assigned to the RAAF's 92 Wing based at Edinburgh, NSW Australia.

NORWAY

The Royal Norwegian Air Force is one of a few nations in the international maritime patrol community to be equipped with modern P-3C Update III Orions. The RNoAF was the first to receive the export version of the Update III from Lockheed in 1989.

The Norwegian P-3Cs are not standard Update III Orions. They have most of the production Update III avionics, including APS-115B radar, ALR-66(v)3 ESM and UYS-1(v) 10 acoustic processor, but have a different communications system and antenna arrangement. This difference is due to the installation of the ARC-207 VHF radio with twice the peak power of standard P-3Cs. The aircraft were also provisioned and equipped with a classified prototype electro-optical system that provides a long range, high resolution visual surveillance capability.

Norway originally purchased five (heavy weight) P-3 Bravos from Lockheed in 1968 to replace aging HU-16 Albatross seaplanes. These aircraft were supplemented by two additional (ex-US Navy) P-3B heavy weight Orions in 1980. Of the seven Bravos, Norway sold five of the aircraft to Spain in 1989 to off-set the cost of the four new P-3C Update IIIs. The two remaining bravos were later modified by Norway specifically for coast guard surveillance duties and civilian taskings.

Norway, after having been neutral during WWII, became a founding member of the North Atlantic Treaty Organization (NATO) in 1949, due in part to its strategic location close to the (now former) Soviet Union. This geographic position related to a strong emphasis on ASW during the Cold War. Soviet surface ships and submarines had to pass through the Norwegian Sea on their way to and from bases on the Kola Peninsula—within 25 miles of the Norwegian territory.

Norwegian P-3C Update III Orion. *Lockheed*

Original Norwegian P-3B. *Lockheed*

The RNoAF's #333 squadron, operating from their island base at Andoya (170 nautical miles north of the Arctic Circle), have prosecuted more Soviet military shipping over the past 30 years than any other squadron in the international maritime patrol community—often detecting several submarines on a single sortie. The squadron's patrol area is seven times the size of Norway itself, extending almost to Greenland over to Iceland, England and Denmark, northward to within 400 nautical miles of the North Pole, over to the border of the Soviet Union and all the waters to the east. Norway is the principle guardian of NATO's northern flank.

The Norwegian P-3's principle mission was ASW during the Cold War, monitoring naval movements within the target rich environments of the Norwegian, North and Berent Seas. Soviet submarines constantly patrolled off the coasts of Europe and North America as well as probing every corner of the Atlantic and adjacent seas. Occasionally, the Norwegian P-3s were directed against suspected submarines located within the numerous fjords scattered along the Scandinavian coastline, as was the case on 1 May 1983. Between 27 April and 6 May, a potentially hostile submarine was detected in the Hardanger-fjord. A Norwegian P-3 dropped three depth charges on the suspected contact deep inside the fjord on May 1st. The next day, another RNoAF P-3 dropped two more depth charges on a contact close to the mouth of the fjord. Norway's #333 squadron is the only P-3 unit in the international MPA community to have ever launched live weapons against a submarine since WWII. .

The Norse P-3's secondary mission is intelligence gathering during Soviet test and exercise activities in the region. This surveillance is a high priority and encompassed the uti-

Norwegian P-3B Orion. ***Baldur Sveinsson***

lization of specialized equipment aboard the Norwegian P-3. One particular piece of equipment tailored specifically for this mission was a side-looking airborne radar. This SLAR radar is of the synthetic aperture type designed for airborne reconnaissance and surveillance. The system was installed onboard two of the RNoAF'S P-3Bs, mounted under the floor—aft of the wings on the starboard side. This included a 10x10 curved honeycomb radome that followed the natural contours of the fuselage. The SLAR modification also incorporates a roll-on/roll-off sensor station (#4) in the starboard side aft observer's station. Although the RNoAF sold off five of their seven P-3Bs in 1989, provisions have been made to mount the SLAR radar system into both their P-3N and Update III Orions.

This mission can be hazardous due to the likely encounter with Soviet interceptors, as was the case on 13 September, 1987, when a Norwegian P-3B suffered in-flight damage when it collided with a Soviet SU-27 Flanker fighter while on a surveillance mission over the Berent Sea. The Orion was intercepted by the Soviet jet, approaching the P-3 on its left

Norwegian P-3C Update III. ***Lockheed***

Norwegian P-3N Orion. ***Terry Taylor***

side. It throttled down to match the P-3's speed and came within six feet of the aircraft. After several erratic maneuvers, the interceptor pulled over onto the portside, behind the right wing. Then it banked left and accelerated under the P-3, thus raising its right verticle stabilizer, which made contact with the #4 engine prop. The engine was feathered immediately and the crew radioed a mayday. The P-3 later landed safely at Banak Air Station, the nearest RNoAF Base. Besides the propeller, the fuselage was riddled with shrapnel battle damage.

Another vital mission that the Norwegian P-3s perform is coast guard surveillance duties—EEZ protection. This tasking was established in 1976 when Norway extended its economic exclusion zone out 200 nautical miles from shore. This move created the need for a long range airborne search and surveillance capability to police the rich zone of natural resources off the Norwegian coast and the surrounding island possessions. The EEZ protection responsibility fell to the newly created Royal Norwegian Coast Guard. But the Coast Guard was immediately burdened by numerous tasks which included maritime search and ocean surveillance—to locate pollution and oil spills, identification of illegal fishing vessels and emergency search & rescue. The RNoAF's #333 squadron was directed to provide regular patrol flights over the EEZ in support of the Coast Guard. This added mission and subsequent increase in Soviet naval activity, prompted the acquisition of two additional P-3 Bravos to supplement the Norwegian P-3 fleet.

The RNoAF Orions, like other P-3 operators, also perform additional mission taskings in support of civilian ministries and scientific institutions. The Norwegian Orions have been utilized for such things as updating area maps, ice reconnaissance, whale migration monitoring and polar bear and seal population census taking (in Arctic regions), as well as

Passenger seats and floor track of the P-3N Orion. ***RNoAF***

occasionally dropping mail and supplies to isolated polar stations or the inhabitants of Norway's territorial possessions, Tiny Bear and Hopen Islands.

After ten years of constant demands placed on the Bravo Orions, the RNoAF began looking to modernize its aircraft. Several proposals were considered, but it was finally decided to replace the outdated P-3Bs with new, more sophisticated P-3C Update IIIs. Under the plan, two of the Norwegian Bravos were to be retained and refurbished specifically for Coast Guard duties. Between June of 1990 and May of 1992, the two aircraft (#154576 and #156603) were flown to NAS Jacksonville and modified by the NADEP there.

Re-designated "P-3N" (the "N" for Norwegian), the specially modified aircraft were tasked with the previously mentioned coast guard duties, as well as crew and personnel transport flights (in connection with simulator training in the United States and in the Netherlands), utilities flights and special operations missions. The Mod program began with the removal of all ASW sensor station equipment (save for the APS-80 radar and AAS-36 IRDS) and electronic/ordinance racks, as well as stripping out of all the sonobuoy chute system and associated plumbing. Installations included upgrading the flight station compatible to the P-3C Update III configuration, facilitating pilot and flight engineer training and adding new floor panels associated with the installation of floor tracks. The tracks run from the aircraft's tactical station all the way back, to accommodate 30 business class passenger seats for the transport of personnel or tying down of cargo on utility missions.

Other features include new navigation aids (dual LTN-72 inertials, ARN-118 TACAN and dual ARN-140 VOR), a new communications suite, a new ESSI radar monitor (for the existing APS-80 radar), new lighting panels and rails for the storage of survival equipment.

Additionally, the aircraft's head was modified to that of airline lavatories and the interior was redone in new (gray) wall coverings and trim. Each of the P-3N were also equipped with a (single) Auxiliary P-3C type, CAD pressurized sonobuoy

Tactical compartment of the P-3N Orion; note the new display monitor at the radar station. ***RNoAF***

Three SAF P-3A Orions. ***Lockheed***

chute and freefall chute. The tubes were installed in the far aft section of the aircraft and gives the P-3N the capability of dropping sonobuoys or smoke markers during SAR missions.

Another little known modification, made to at least one of the P-3Ns, was the installation of an electro-optical surveillance system known as "Cluster Ranger." This EOS system is a standoff, high-resolution optical surveillance system used for long range visual surveillance. The Cluster Ranger system was installed in the aircraft's old radio operator's station (dubbed sensor #5) on the forward port side.

Since 1992, the RNoAF has been developing a P-3 upgrade program to enhance the ASUW capability of its four P-3C Update III Orions, while improving the ASW capacity of each aircraft. Known as the "Upgrade Improvement Program" (UIP), the proposed upgrade program includes installation of the CP-2044 tactical computer produced by UNYSIS (a number of units have already been acquired and are being stored until the project gets underway), adding an imaging search radar, GPS navigations, SATCOM communications and the incorporation of integrated displays. The UIP program would also upgrade the aircrafts' ESM system, which includes the establishment of a dedicated operator position solely for the management of the ESM system. Other components of the proposed upgrade include the addition of self-defense system provisions comprising a radar/laser missile warning system and chaff/flare dispensers. These improvements are consistent with those being made by the US Navy and other Orion operators.

SPAIN

The Spanish Air Force has been an Orion operator since it acquired three ex-US Navy DELTIC P-3A aircraft in 1973 to supplement its fleet of HU-16 ASW equipped Albatross seaplanes. Later in 1979, due to the crash of one P-3A in 1977 and the subsequent retirement of the Albatross from service

Spanish Air Force P-3A Orion. ***Eric van Rossum via Marco Borst***

in 1978, the SAF entered into a ten year lease agreement with the US Navy for four DELTIC P-3As. During the Cold War, the Spanish Orions were a vital link in the protection of NATO's southern flank. The P-3 guarded the natural choke point between the Atlantic and Mediterranean at Gibraltar. Other missions included patrolling the vast open ocean of the Atlantic, monitoring Soviet naval activity transiting south towards points along the African Ivory Coast.

Since the end of the Cold War, the SAF Orions have continued to play an important role in NATO's Maritime Patrol community. Missions still include ASW, as well as EEZ protection, coastal patrol and maritime surveillance. Current operations include deployments to Son Sanjuan Air Base in the Balearic Islands, Gando Air Base in the Canaries and Santiago de Compostela Air Facility northwest of the Iberian Peninsula.

In recent years, the SAF has added UN/NATO embargo enforcement to its list of responsibilities having participated in the Operation "Sharp Guard," the UN imposed embargo against the former Yugoslavian republics of Bosnia and Herzogovina. The Spanish P-3s are also regular participants in NATO exercises both in the Atlantic and the Mediterranean.

Back in the late 1980s, as the US Navy's P-3 Alpha lease with Spain was nearing the end of its term, the SAF made a request for six new P-3C Orions, but the Spanish Govern-

Spanish P-3B in new tactical paint scheme and markings. ***SAF***

Spanish Orions on the flight line at Jerez air base. ***SAF via VPI***

ment couldn't afford the expense of new aircraft and options to buy used P-3A and or P-3B models were explored. Finally, a decision was made to buy five (heavy weight) P-3 Bravos from Norway, which were being sold to off set the cost of new P-3C Update III. Although these were standard DELTIC P-3 Bravos, the aircraft did have some improvements made to them while with RNoAF that encompassed the AQA-7(v)2 acoustic processor and AQH-4(v)2 acoustic tape recorder, LTN-72 inertial navigation set, AAS-36 IRDS and a ARN-118 TACAN nav-aid which are indicative of a TACNAVMOD (Block 2) aircraft—but lacks the general purpose digital computer of that configuration. The five P-3Bs were acquired from Norway between April 1988 and June 1989.

In 1990, it was planned that the aircraft would be upgraded. This upgrade was to have involved improvements made to the radar, acoustic suite and the ESM system, while adding a FLIR and a Harpoon missile capability. The aircraft were also to have had a new navigation and communications suite installed along with new integrated displays. The Upgrade program was to have utilized Spanish industry and identified particular equipment, such as a (UK) GEC Avionics digital mission computer and AQS-934 or -948 acoustic signal processor. Other systems were to have included the Thorn EMI Searchwater II (multi-mode) radar, an ALR-66(v)3 ESM system and FLIR Systems, Inc. Series 2000 forward looking thermal imaging system. Unfortunately, Spain slashed its military budget in early 1992 (to help pay for hosting the 1992 summer Olympics) and the upgrade program was discontinued.

The aircraft were subsequently sent through standard depot level maintenance (SDLM) at OGMA, the Oficinas Gerais de Material Aeronautico aircraft rework facility at Alverca AB in Portugal. Upon completion, the SAF Orions were placed in operational service. This SDLM included a new tactical gray paint scheme that deviated from Spain's traditional gray and white maritime scheme of the P-3A. After the leased P-3As were returned to the Untied States, the remaining two Spanish P-3As were re-engined with T56-A-14 turboprops to provide better performance and to streamline maintenance compatibility.

Up until 1994, the Spanish Orions had been operated with #221 Squadron (#22 Wing) based at Jerez de la Frontera AB near Cadiz, just a short distance from Rota, Spain. Now the Squadron is flying the Orions from their new air base at Moron.

Although the P-3 Orions are owned and maintained by the Air Force and have Air Force pilots, the tactical crew is made up of naval personnel. Hence, the operational control

Spanish Orion turning toward the future. ***SAF***

Iranian P-3F Orion. ***Lockheed***

of the squadron and the aircraft is held by the Spanish Navy's Commander of the Fleet, a relatively new function created in January 1988.

As for the future, the SAF is still trying to initiate some kind of an upgrade program. The current proposed plan has suggested addressing each upgrade change individually as funding becomes available and identifies those systems from the previous upgrade program. This one-at-a-time approach will allow Spain to enhance its Orions to fulfill both its national and international MPA requirements.

IRAN

In the continuing development of the P-3 Orion by Lockheed, there have been many firsts and many derivative Orion aircraft produced. But when the middle eastern nation of Iran placed an order for six P-3C Orions in early 1973, little did anyone know then that these six aircraft would go on to establish a number of production firsts, including a new derivative variant model and later generate one of the most interesting stories about the P-3 Orion, full of twists and intrigue.

It all began back in 1973 when the Imperial Iranian Air Force made a request for six P-3C Orions and became the first foreign (export) operator of the P-3C. The IIAF was in the middle of a modernization program to equip its air force with the most modern aircraft available. Besides its Orions, the IIAF had also ordered and subsequently received high tech aircraft such as F-14A Tomcats, F-4 Phantom attack planes and Boeing 707 air-to-air refueling tankers, as well as various types of helicopters.

In the beginning, the Iranian Orions were to have been utilized for typical maritime patrol operations such as long range surface surveillance and ASW. But right from the start, the US Navy refused to let Iran have the sophisticated systems of the P-3C and any of those systems being developed for the Update I program. There was some discussion for a while that the Orion aircraft would be delivered without avionics. But in the end, the decision was made to provide the aircraft with the search capabilities of the P-3A/B Orion—save for the inclusion of an APS-115B search radar, LTN-51 inertial navigation system, ARN-84 TACAN and APN-153 DOPPLER navaids, as well as a surveillance camera fit common to the P-3C NUD. The airframe itself was a P-3C on the outside, including the underbelly sonobuoy chutes, but the interior arrangement was more a combination of the Alpha/Bravo and the Charlie layouts. There was the P-3A/B side-by-side tactical workstations for the acoustic sensor operator, TACCO and sensor #3 operator with the navigator's position located in the P-3C's forward TACCO position (on the portside) and the radio operator-communications suite across the aisle on the starboard side. This configuration, with its differences in avionics and airframe, combined to demand that a new des-

P-3F over maritime environment. ***Lockheed***

Iranian P-3F's lined up at Moffett Field in original gray and white paint scheme and markings; note the first camouflaged painted P-3F in line. ***Lockheed***

ignation model/type be established. The next letter designation available on the model/class list was "F," hence the "P-3F" Orion was born.

There has been a rumor associated with the P-3F over the years that has suggested that the aircraft had an in-flight refueling receptacle installed as a production feature. The in-flight refueling capability was to have given the Iranian Orions a very long range once refueled by Boeing 707-3J9C tankers that were also purchased by the IIAF. Despite previously published accounts, the P-3F Orions were never equipped with an IFR capability. The US Navy didn't even begin to conduct in-flight refueling proximity tests on the Orion until 1979, let alone have this capability installed in Iranian P-3Fs in 1974-75. There may have been some discussion on the matter during the program or potentially initial requirements established by the IIAF, but the fact remains that the aircraft were not equipped with IFR.

P-3F Orion ***B. Stewart via Isham collection***

As a new P-3 operator, Iran's Orion operational capability was marginal. It was the country's strategic location, right in the center of the Middle East, that was important. Iran is bordered on the west by Iraq and Turkey, as well as Saudi Arabia across the Persian Gulf and Oman separated by the Gulf of Oman—to the North, the Soviet Union, with Afghanistan and Pakistan to the east. To the south lies the Arabian Sea and the vastness of the Indian Ocean.

It was Iran's strategic location that would provide the US Navy with a new operational Orion compatible base in the Middle East from which to launch operations into the Persian Gulf, Arabian Sea and Indian Ocean. It wasn't too long after the first P-3Fs arrived in country, that the tempo of US Navy Orions showing up on the ramp at Bandar Abbas (where the P-3F initially operated from) increased.

The Iranian Orion program was also important to Lockheed, because they had been selected to provide flight training, ground maintenance training and logistical services to the program in country—so important that it sent a contingent of employees over to live and work in Iran. Approximately 80 Lockheed instructors, technicians and maintenance personnel (some with families) moved to Bandar Abbas (airport) airbase. It became the largest and most extensive support effort with a foreign customer/aircraft sale that the company ever undertook. There was also an early indication that more Orion aircraft were desired by the IIAF and was probably the cause for the extent of the support program.

By February 1975, the first P-3Fs began to be delivered to NAS Moffett Field where VP-31 was in the process of training Iranian Orion crews. There had been no MPA units previously in the IIAF and a new squadron was going to be formed as a result the Orion purchase. It was to have been made up of air force pilots to fly the planes with the Iranian Navy personnel operating the tactical suite. But in the end, the P-3F's compliment was made up of an all air force crew. In fact, the P-3F's crew were all officers. Besides the officers that piloted the aircraft and filled the TACCO and NAVCOM positions, there was a cadre of "specialist officers" known as "HOMAFARS" that were responsible for radar, MAD, ESM and acoustics management. They ranged from junior (warrant) officers to senior officers, including majors and colonels.

It was at Moffett field that another characteristic of the P-3F was incorporated. The aircraft were originally delivered in the traditional gray and white maritime patrol paint scheme with Iranian Air Force markings. The P-3F were being flown by VP-31 and Iranian aircrews as part of their transition training. During an inspection of the training operations by a visiting high-ranking Iranian General, the General observed a taxiing USN A-4 Skyhawk jet painted in a unique eastern block camouflage paint scheme (the A-4 was an aggressor aircraft utilized by the Navy's TOPGUN school). The Iranian General was so taken with the unusual camo pattern that he ordered all the P-3Fs to be immediately painted similarly. This became a problem because the P-3Fs were used almost daily for pilot training. Lockheed finally worked it out, sending an aircraft (one at a time) to its Ontario, California, repair facility to be painted over consecutive weekends. So, by the time the P-3Fs were delivered to Iran in April of 1975, all the aircraft were sporting the distinctive new camo paint scheme that has become synonymous with the P-3F. The camo paint scheme, being very striking, was also very effective. During the P-3F's training flights at NAS Moffett Field, the control tower's traffic controllers had a difficult time maintaining visual contact with the P-3Fs within the air field's traffic pattern. Constant radio contact between the aircraft and the tower was necessary in order for the pilot to communicate the aircraft's movements within the circuit and during runway approaches and landings.

After the Orion's delivery to Iran in April 1975, the Lockheed training schedule began again, punctuated by many training flights conducted over the vast desert interior of Iran and those flown over close-in coastal waters. But occasionally, the aircraft would be utilized operationally, taking off and heading out over the Arabian Sea and southeast over the Gulf of Oman or northwest across the Persian Gulf. These were mysterious flights, occasionally without Lockheed instructors onboard. Once the training program ended in January 1979, flights throughout the region were conducted, including those flown south out into the Indian Ocean.

As the first years passed (1975-1977), the IIAF began negotiations toward the purchase of 12 more P-3C baseline airframes to increase its Orion fleet to 18 P-3Fs. Three aircraft were actually ordered during this time frame, but tensions within Iran began to grow. Iran was a predominantly traditional Moslem state and the Shah (imperial leader of the country) had been making many western style changes. Religious factions began to gain more and more power. By the fall of 1979, the situation came to a head with the instigation of an Islamic Revolution to overthrow the Shah of Iran. The Shah fled the country and the Islamic Revolutionary council took over governing the country. Relations between Iran and the US soon broke down, with the eventual taking of American hostages at the American Embassy in Tehran.

A rare interception of a Iranian P-3F Orion. *US Navy*

During this time, all contact with the Iranian Orions was lost. A series of economic and military embargoes were implemented, isolating Iran from the logistical supply of parts to keep the P-3F flying. It seems now that many changes occurred during this period. The IIAF gave way to become the (Iranian) Islamic Revolutionary Air Force and included the moving of P-3F operations from Bandar Abbas (tactical air base #9) to the tactical air base #7 at "Shiraz."

It also seems that it was there at Shiraz on February 15, 1985 that a P-3F crashed. This particular P-3F, aircraft #159343, had the distinction of being the only P-3F to have been modified by the US Navy and Lockheed with a Harpoon anti-ship missile capability. The aircraft had undergone Harpoon weapons tests at the Pacific Missile Test Center (PMTC), NAS Point Mugu, California, during 1978. The P-3F returned to Iran with a number of Harpoon missiles on board just prior to the Islamic Revolution.

This crash decreased the number of available, flyable P-3F Orions to five. For the next several years (through the late 1980s) no fewer than one or two P-3Fs were ever seen flying at any one time. The belief was that the other aircraft were used as parts aircraft to supply the flyable ones. But recent intelligence reports have now indicated that after 1988, more P-3Fs were seen flying during the war between Iran and Iraq. It's further believed that a source of spare parts was possibly provided during the Iran/Contra incident. This is where mili-

Dutch (Netherlands) P-3 Update II.5 Orion. ***Lockheed***

tary weapons were secretly sold to Iran by the US Government, despite international embargoes against it. It seems that aircraft parts were included in the military equipment sold to Iran and that some of those aircraft parts included spares for the P-3F Orions. During the more recent Gulf War with Iraq (1990-91), two Iranian P-3F Orions were seen flying in the Persian Gulf. Only one has been seen flying since then.

It now seems that the end of P-3F operations may be at hand. As recent as 1993, Iran bought five Dornier maritime patrol aircraft from Germany. There is also information to suggest that Iran plans to produce more of these aircraft for the Iranian IRAF domestically in its aircraft production facilities. If so, the remaining P-3Fs could likely be withdrawn from service and thus bring to an end twenty years of venerable service.

NETHERLANDS

The Royal Netherlands Navy became the first European customer to order the P-3C Update II.5 when it placed a request for thirteen Orion aircraft in December of 1978. These Orions are not standard Update II.5 aircraft. The RNLN actually ordered the current production aircraft available at the time, the P-3C Update II, but received the aircraft lacking several of the components characteristic of the Update II configuration, including the IRDS (and its associated recorder) and Harpoon missile control system. The aircraft were provisioned for these systems, but were just not equipped with them. Also, as the Netherlands P-3 program progressed, Lockheed introduced Update II.5 changes into production. So, in the end, the aircraft were delivered and designated as P-3C Update II.5 Orions.

The RNLN's aviation service, known as Marine Luchtvaartdienst, or MLD, had been interested in the P-3 prior to its placing an order in 1978. As far back as 1968, the MLD operated SP-2H Neptunes and carrier-based S-2F Trackers. The Dutch Navy was in the process of retiring its only aircraft carrier, the HrMs Karel Doorman, and replacing its S-2 Trackers with land-based aircraft. The MLD wanted to purchase the new P-3B Orion, but the Dutch government selected the NATO favored Breguet 1150 Atlantic maritime patrol aircraft. Later in 1974, the government was once again looking for a new maritime patrol aircraft to replace its remaining fleet of P-2 Neptunes and began studying the available MPA aircraft. They reviewed the capabilities of the British Nimrod, the newly proposed Atlantique II and the latest version of the P-3C. They finally selected the Orion.

The Netherlands' geographical position on the North Sea equates to maritime surveillance as the Dutch Orion's primary mission. The area is congested with merchant traffic heading for the busy commercial seaports of Great Britain (just a short way across the North Sea) and the countless fishing fleets departing ports along the coasts of England, Scotland, France/Belgium, Germany, Denmark and Norway. Little has changed since the Cold War, except for the decrease in Soviet surface and sub-surface military activity. Then, the Dutch P-3 (along with Norway's Orions) were the vanguard against the red fleet, protecting NATO's northern flank.

In more recent years, new missions have arisen to keep the Dutch P-3s busy. Since 1991, the Dutch Orions have taken over airborne support duties for Netherlands Coast Guard. This tasking is due in part to the coast guard losing the lease on its patrol aircraft and the need for the RNLN's P-3 to find new missions. The new coast guard commitment includes fisheries patrols, search and rescue missions, pollution monitoring, environmental surveillance and counter narcotics flights over the Dutch territorial waters of the North Sea. These flights usually include a Dutch Police officer on board and consist of gathering evidence on suspected violators that can be used in international maritime courts.

Other similar missions encompass deployments to Dutch territories such as the Islands of Curacao and Surinam in the western Caribbean, off the coast of Venezuela. Here, Dutch P-3s have been participating in the war on drugs, conducting counter narcotics operations out of Hato airfield on the Dutch Antilles island of Curacao. Originally begun as a seventeen-day detachment during specific times of the year, the flights out of Curacao have now become a permanent deployment with rotating crews and aircraft every few months. Another Caribbean Dutch P-3 mission tasking includes operations in economic protection support of the island nation of Surinam, a Dutch Protectorate. This mission encompasses fisheries patrols to detect illegal fishing vessels within Surinam's EEZ and territorial waters. The island nation was offered the Orions by the Dutch Government since Surinam has no suitable aircraft of its own to conduct such missions. These P-3 flights, also flown out of Hato airport on Curacao, are conducted in close cooperation with the Surinam National Army, Navy and Customs Police.

Besides missions of national interest, the Dutch P-3s play an active role in NATO operations worldwide. During the recent Gulf War with Iraq (1990-91), two Dutch Orions took over NATO responsibilities of US Navy P-3s in the Mediterranean. Flying out of NAS Sigonella, the Dutch P-3s freed the US Navy Orions to participate in Gulf War operations and support coalition forces.

Other RNLN Orions worked closely with Dutch frigates during Gulf War embargo enforcement operations, while another aircraft supported the Dutch military hospital located in Jebel Ali, United Arab Emirates. This included one aircraft (#161374) modified with a medical transport configuration consisting of eight stretcher-beds, medical equipment and eight additional passenger seats to evacuate possible casualties of the Gulf War.

Since 1992, two Dutch P-3s have been permanently detached to NAS Sigonella for participation in the United Nation's Operation Sharp Guard, the imposed trade embargo against the former republics of Yugoslavia. This includes surface surveillance operations over the Adriatic Sea with live torpedoes stowed in the aircraft's weapons bay.

In a similar mission to Sharp Guard, Dutch P-3s have participated in the NATO enforced embargo against the island of Haiti in the Caribbean. Known as Operation Support Democracy, the Dutch Orions flew surface surveillance missions in support of Dutch Frigates. The aircraft were based at Curacao and also conducted counter narcotics flights as an extension of their Curacao operations in support of Surinam.

Another international commitment that the Dutch Orions support is Iceland. The RNLN contributes a one plane detachment to NAS Keflavik operations per a military agreement made at the request of the Icelandic government to have the aircraft permanently deployed there. The aircraft and crew falls under the command of NATO, but operates as part of the Iceland sector ASW group in unison with the US Navy.

Dutch P-3C Update II.5 in current markings. ***Royal Dutch Navy via VPI***

Dutch P-3 over submarine *Royal Dutch Navy via VPI*

One Dutch P-3 Orion is permanently stationed in Iceland at the request of the Icelandic Government. ***Baldur Sveinsson***

The Dutch P-3s conduct ASW training operations right alongside the American Orions to support the primary NATO mission of maintaining a presence in the Arctic as well as participating in NATO interoperability operations such as mine warfare exercises. The training benefits to the Dutch crews are immense.

The Icelandic detached aircraft is also positioned to provide vital support to their own air force. This includes assuming the SAR Guard for transiting Royal Netherlands Air Force F-16s from bases in Europe to joint exercises in Canada. Known as the "Goose Bay Ferry," the Orions perform communications relays and maintain a SAR-ready status should one of the F-16s experience trouble over the water and the pilot has to bailout.

Since 1993, the Royal Netherlands Navy has been developing an upgrade program to modernize their fleet P-3C Update II.5 Orions and improve the aircraft's anti-surface warfare (ASUW) capability. They have been conducting a study for several years now to establish requirements for the Capabilities Upgrade Program, or CUP. Several years ago, shortly after the demise of the Soviet Union and the end of the Cold War, the RNLN was considering a plan to mothball four or five of the P-3s due to a perceived lack of missions and a restructuring plan to reduce the operational organization by one squadron. There were also proposed plans suggesting selling off a third of the P-3 force (Germany was mentioned as a potential customer) and sell or trade in the whole fleet towards the purchase of a smaller number of more advanced aircraft like the then proposed Lockheed Orion II. It should be noted that during this time (1991) the Dutch P-3s were the lowest hour P-3C Update II.5 Orions in service world wide.

But in the end, the decision to upgrade the existing fleet was determined. By 1993, the CUP program was identified and promised to afford the RNLN commonality of aircraft capabilities with those of other Orion operators.

Key areas for improvement under the CUP program included a new central processor/data management system, adding an imaging radar, upgrading the ESM system and the installation of a new acoustic system with an associated 99 channel sonobuoy receiver set replacing the existing AQA-7(v)8. The CUP also plans for the installation of multi-purpose, color, high-resolution displays. Proposed systems may comprise the Unysis CP-2044 computer, the APS-137 ISAR radar, ALR-66(v)3 ESM system or Australia's ODYSSEY 2000 ESM system and UYS-503 acoustic system. The CUP program could begin as soon as 1997 with a prototype delivered by the turn of the century and production commencing by 2002.

In the meantime, the Dutch are proceeding with a number of interim avionics upgrades which are essential for the aircraft to meet a number of their international mission requirements. They encompass new UHF/VHF radios for the communications suite, GPS set for navigation and the installation of a new forward looking infrared/thermal imaging system by FLIR Systems, Inc. The AAQ-22 SAFIRE thermal imaging system is a gyro-stabilized, high-resolution digital thermal imager. The FLIR systems were originally acquired for Dutch H-14 Lynx helicopters, but found their way into several of the RNLN P-3. The system fits the existing IRDS-provisioned retractor mechanism and sensor-operator console. The thermal imaging system has greatly enhanced the Dutch P-3's surface surveillance mission capabilities in the Adriatic (in support of Operation Sharp Guard) and counter narcotics operations in the Caribbean.

The RNLN has also recently revamped its P-3 Orion scheduled depot level maintenance (SDLM) program by awarding the 13.6 million dollar P-3 overhaul and rework contract to OGMA, the Portuguese government's Oficinas Gerais de Material Aeronautico. OGMA began conducting SDLM rework on the Dutch P-3s starting in 1994, cycling them through Alverca, Portugal. This contract was previously held by KLM, the Netherlands airline.

CANADA

Of all the P-3 Orion variants produced over the years, the Canadian version has been the most unique, with characteristics that have now proved to be ahead of their time. Not just because of the aircraft's unique configuration and capabilities tailored specifically for Canada's diverse requirements, but because of how those unique elements are now affecting the next generation of maritime patrol aircraft in the future.

The Canadian CP-140 Aurora was developed as a long range surveillance aircraft providing the Canadian Forces not only with an efficient submarine hunter, but also a multi-mission platform with capabilities to conduct a multitude of national interest taskings required by the Canadian Government.

The CP-140 shares the same airframe and engines of the highly successful and flexible Lockheed P-3C (Update II), but contains the advanced systems and avionics of the

Canadian CP-140 Aurora. ***CF /DND***

equally-successful Lockheed S-3A Viking modified for the Aurora. The principle difference of the Aurora is its interior tactical crew arrangement, clustered in the center of the aircraft. This unique "horseshoe" or "U" shape cluster design fosters improved crew coordination and mission efficiency among the sensor operators.

Like the P-3 Orion, the Aurora was primarily utilized as a submarine hunter, a vital link in NATO's defense of the Arctic and North Atlantic during the Cold War. The US Navy benefited from Canada's Aurora acquisition through increased ASW coverage of North Atlantic and Arctic regions with two new P-3 service equipped bases in the far north, one in Greenwood, Nova Scotia, and the another in Comox, British Columbia.

Development of the Aurora can be traced back to February 1969, when the new unified Canadian (Armed) Forces announced that a replacement for the current maritime patrol aircraft, the CL-28 Argus, was needed to counter the growing Soviet submarine threat. But initiation of the replacement program was considered a low priority among other acquisitions by the government and was delayed until August 1971 when the Long Range Patrol Aircraft management board was formed. The LRPA management group was set up to establish requirements for the new Canadian patrol aircraft. Later on 20 July, the LRPA board announced that it would be accepting request for proposals (RFP) from Industry.

Of the proposals received for LRPA consideration, one offered a Canadian version of the Hawker Siddeley Nimrod, while others proposed the Dassault Brequet Atlantic, a McDonnell Douglas DC-10 and a Boeing 707 modified to a ASW/MPA configuration. Originally, Lockheed proposed three varied aircraft proposals based on the P-3 Orion. They ranged from a standard US Navy P-C Update I /II, to a new derivative aircraft tailored specifically for Canadian requirements. There was even one proposal suggesting a refurbishment of the then current Argus airframes. This included re-engining the Argus with turboprops and refitting it with new state-of-the-art avionics. The field of competitors was later reduced down to two aircraft proposals, the Boeing 707 and the Lockheed P-3C Orion. The winning bid hinged upon which

Prototype CP-140 Aurora in original markings. ***Lockheed***

CP-140 exterior ; underside features. ***CF /DND***

company would spend the most money in Canada through sub-contracted work. Lockheed's P-3 LRPA won the contract, with a number of Canadian firms contributing components to all P-3 production over the next fifteen years.

An order for eighteen aircraft was placed by the CAF with Lockheed on 27 November 1975. The Eighteen LRPA aircraft replaced thirty-three Argus. The lower number of replacement aircraft was determined based upon the proposed capabilities of the LRPA platform, including the aircraft's higher speed, advanced avionics and quicker maintenance turnaround. The lower number of LRPA aircraft were deemed capable of handling the workload of twice as many Argus. Considering the size of Canada's population, its geographical size and state of its economy, the purchase of eighteen aircraft is equivalent to a 200 aircraft order in the United States.

After awarding the contract to Lockheed, Canada sent (air) forces personnel to Lockheed in California to form a design team. The resulting aircraft that developed became the CP-140 Aurora. The Aurora's airframe, as previously stated, is based on the P-3C Update II. The aircraft's tactical suite is built around the multi-functional avionics of the S-3A Viking and encompasses its APS-116 search radar—capable of detecting small targets (Canadian designation APS-506), the OL-82 acoustic processor (Canadian designation OL-5004) and its associated ARR-76 sonobuoy receiver/ARS- 501 sonobuoy reference system. The Aurora aircraft also utilizes the S-3's AYK-10 digital central data processing computer (AYK-502) and entry keyset controls and tactical displays. The computer integrates all onboard functions including management of sonobuoy activities and weapons stores. It's also capable of performing hands-off flight to TACCO calculated fly-to-points for automatic release of sonobuoys. The computer also maintains the aircraft's threat library to assist in classification of unknown contacts and performs all flight and mission record keeping, storing the data on digital magnetic tape for later analysis.

There are also a number of features of the Aurora that are unique to it and not common with either the Orion or the Viking. They include the pilot's ordnance control panel, the cockpit's auto flight control system, the aircraft's communication and navigation gear and its new MAD system. The ASQ-502 MAD, developed in Canada by CAE Electronics, Ltd., is an advanced integrated MAD system containing an auto-compensating sub-system. The aircraft also utilizes a domestically produced ESM system tailored to Canada's unique maritime environment. The system is a quick, omni-directional electromagnetic signal detection system with a threat alert feature and automatic classification of known radio and radar emitters. The ESM system, similar to the S-3, is unique to the Aurora with its antennas mounted on the aircraft's wingtips. Even the aircraft's ARN-511 OMEGA and APN-510 Doppler navaid sets are different.

Another unique feature aboard the Aurora is the Zeiss KS-501A camera system. This is a high resolution vertical panoramic camera system that permits day or night (passive) photography through the use of infrared film with an associated infrared illuminator strobe.

CP-140 unique "U" configured tactical compartment. ***Lockheed***

CP-140 tactical compartment. ***Lockheed***

Aurora over mountain terrain. ***CF /DND***

The aircraft also possesses 36 belly-mounted CAD launched sonobuoy chutes verses the 48 of the P-3C and the 64 of the S-3A. The space saved in the airframe from having less sonobuoy chutes provides the room for positioning the Zeiss camera system.

Besides the difference in the interior tactical compartment, the Aurora is also unique for re-introducing the forward observer stations on the port and starboard side of the aircraft. The Aurora also has a redesigned crew rest/galley area in the aft section of the aircraft. The Aurora is also equipped with the USH-502, a commercial type flight data recorder or "Black Box" that records voice and flight data in the event of a crash.

The Aurora also has the distinction of establishing avionics and systems onboard the Orion airframe that would later be considered by the US Navy and subsequently incorporated into USN Orion production. These systems include an improved environmental control system and the FLIR infrared sensor.

The Aurora is also unique in that it was originally provisioned to carry special pelletized instrument packages mounted in the weapons bay for civil national interest taskings. The weapons bay was equipped with plug-in power and cooling outlets to supply the pelletized packages. One such weapons bay mounted system that has been tested on the Aurora was the APD-10 side-looking airborne radar.

Aurora over maritime environment. ***CF /DND***

CP-140 Arctic operations. ***CF / DND via VPI***

Original CP-140. ***Lockheed***

In more recent years, the Aurora has been equipped with a weather reconnaissance radar system. The WX-1000 Stormscope is a commercial color weather radar providing for weather avoidance and was installed in the Aurora fleet through early 1996.

The LRPA/Aurora program also included the establishment of a new data interpretation and analysis center at CFB Greenwood. Housed in a newly constructed building on base, the DIAC was accompanied by an operational mission and flight simulators.

With the Pacific Ocean to the west, the Atlantic Ocean to the east and the Arctic to the north, Canada possesses one of the world's longest and most unusual coastlines with over 36,356 miles of coastal mainland and 115,133 miles of islands. To maintain a constant vigil over this vast and rather inaccessible land requires patrol by aircraft. The Canadian Aurora is like the vast and varied territories that it patrols. The aircraft was designed with capabilities for various national interest roles including natural resources surveillance, environmental monitoring, fisheries protection, law enforcement support and northern Arctic sovereignty patrol.

Many of the civilian tasks resulted from the Department of National Defense (DND) signing Memorandums of Understanding (MOU) with various Canadian government agencies to provide them with aerial support. Now the Aurora's multi-mission capabilities are being fully utilized.

Environmental Canada utilizes the Auroras for monitoring of pollution and control of natural resources, which includes animal population census taking and aerial mapping. Some of these missions are accomplished through design

Aurora Adriatic operations. ***CF /DND via VPI***

Aurora taking off towards the future. ***Baldur Sveinsson***

provisions in the aircraft and with the specialized weapons bay mounted packages to house sensors and photographic equipment.

In support of the Department of Fisheries and Ocean's enforcement program, Canadian Auroras monitor shipping traffic in maritime regions to locate and identify violators of Canada's 200 mile EEZ. Auroras investigate suspected vessels, querying them as to their point of origin, cargo on board and amounts and/or types of fish collected. Violators are identified and photographic evidence is generated to aid in prosecution. Canada is also a member of the North Atlantic Fisheries Organization (NAFO) and is responsible for policing fishing zones outside its 200 mile limit.

Through a MOU between the DND and the Royal Canadian Mounted Police (in association with Canadian Customs), military Auroras provide aerial surveillance to monitor and track surface vessels suspected of smuggling drugs into maritime provinces. These missions usually include drug liaison officers from the RCMP aboard the aircraft to coordinate operations. The Auroras locate, identify and track these ships until seagoing law enforcement or Coast Guard assets can intercept and board the craft. If contraband is found and arrests are made, photographs and data collected by the Aurora are used as evidence in court. Canadian Auroras have been involved in many such operations over the years, with several recent captures netting very large quantities of drugs bound for Canadian cities. Another important national interest mission is northern sovereignty patrols of Canada's Arctic region. This mission stems from a 1985 government mandate for the military to demonstrate a presence in isolated Arctic and coastal territories, to assert Canada's sovereignty and protect its natural resources from those who would exploit them.

An important sub-mission of northern sovereignty patrols is Ice Reconnaissance. Icebergs located during these NORPAT flights are reported to the Iceberg Central Division of the Canadian Department of Transportation. The DOT "Ice Central" uses iceberg location data in its maritime radio reports to mariners for safer navigation of northern waters.

Other northern patrol taskings include pollution monitoring, environmental surveillance, emergency medical evacuation and Arctic search and rescue. For this mission the Aurora are designed to carry a variety of auxiliary equipment including the survival kit air droppable system, or SKAD. The SKAD is a multi-purpose life raft and survival equipment canister used during search and rescue (SAR) missions. The survival package was designed to fit the Aurora's weapons bay and provides a significant measure of emergency survival equipment for any search and rescue mission.

The first Aurora rolled out of the Lockheed Burbank production facility on January 25, 1979. Its maiden flight took place on 22 March. Over the next year, Canadian forces instructors underwent training at Lockheed, culminating in the delivery of the first aircraft to CFB Greenwood, Nova Scotia, on 29 May, 1980. It was there during November 1980 that Maritime Patrol Squadron 405 became the first Canadian operational unit to be equipped with the CP-140 Aurora. The squadron conducted its first operational mission of the Canadian Aurora during March 1981.

Although expanding national interest roles of the Auroras are important in the Post Cold War era, traditional military roles of the Aurora are still in demand as demonstrated during the recent support of NATO operations in the Adriatic. The Canadian Auroras had been participating in operations to support United Nation's resolution 820, enacting an arms embargo against the former republics of Yugoslavia. As part of Canada's NATO commitment, beginning in September 1993, the Auroras flew armed surveillance missions over the Adriatic from the naval air station at Sigonella, Sicily. Known as Operation Sharp Guard, the support missions included surface surveillance operations to detect, identify and query any merchant ships in the region. Those that were suspected of carrying prohibited cargos to the former Yugoslavian republics were reported to the fleet aircraft command and control ships. Suspect vessels were then targeted for interception and boarded for inspection by other NATO surface units. If found to be violating the imposed embargo, the ships were escorted to port for discipline. Operation Sharp Guard was one of the first major international deployments for Canadian Aurora squadrons.

In the meantime, Canada has not been immune to budget cuts and defense downsizing. Like the US Navy, Canada continues to maintain its domestic mission taskings and obligations to NATO in an environment of dwindling budgets and funding reductions. Despite the cuts, the DND recognizes that the Auroras are in need of upgrading to maintain the operational capabilities of the aircraft in the future. In an effort to ensure the operational effectiveness of the Auroras (till 2010) and maintain its capabilities for interoperability with other NATO forces around the world, the Canadian Forces have initiated the Aurora Life Extension Project, or ALEP. The project was established in 1992 to address numerous operational deficiencies and equipment obsolescence, as well as supportability issues that have developed in the Aurora over the last eighteen years.

ALEP equipment replacements will emphasize commercial-off-the-shelf (COTS) and non-developmental systems with some equipment having a potential commonality with the Canadian maritime helicopter project. The scope of the ALEP program includes replacement, refurbishment or acquiring lifetime spares that are currently insupportable until 2010.

The program also seeks to restore declining capabilities while improving existing surface surveillance, interoperabilitiy and reliability capabilities of the Aurora. While still undergoing definition of requirements and identification of specific systems (early 1996), some components of the upgrade project include a new imaging radar.

Under ALEP, a new Spotlight Synthetic Aperture Radar (SSAR) modification to the existing APS-506 radar is to be conducted. Currently being tested, the Spotlight SAR Mod adds four new radar modes to those of the Aurora's existing APS-506, to give the aircraft long range stand-off surveillance capabilities. The different radar modes include both land and sea Spotlight SAR modes, a range-Doppler profiler mode (essentially an enhanced ISAR mode) and a strip map mode. The SSAR radar modification is a significant modification that, once completed, will encompass a re-designation of the radar unit consistent with its uniqueness and the Aurora itself. There is also an interest to integrate a new IFF interrogator to the new radar modification.

Other ALEP upgrades include a new acoustic suite. The Aurora's OL-5004 acoustic processor, ARR-76 sonobuoy receiver and ARS-501 sonobuoy reference system are fast becoming insupportable and no longer effective against the modern quiet nuclear and advanced diesel submarines. The new acoustic processor is to be a COTS system with thirty-two channels providing simultaneous processing of narrowband, broadband and transient signals. The system is required to possess built in growth with embedded training and the potential for multi-static low frequency active operation, array processing and expansion to 64 channels in the future.

In regards to the Aurora's sonobuoy receiver set and sonobuoy reference system, the ALEP project has selected the Flightline, Inc. ARR-502A system which combines the functionality of the two systems it replaces. The ARR-502A

Aurora in post flight wash-down. ***CF /DND via VPI***

CP-140A Arcturus. ***CF /DND***

permits monitoring of up to thirty-two sonobuoy channels over the ninety-nine channel RF spectrum. The system will utilize the aircrafts' existing SRS antennas configuration—save for the installation of additional blades. The new system provides the Aurora operation of all the current and future sonobuoys and an ASW interoperability with the rest of the international MPA community.

The Aurora's MAD is also scheduled to be replaced due to supportability problems. Requirements mandate the system have equivalent capabilities. The same goes for the Aurora's ALR-502 ESM system. The ALR-502 is also experiencing maintainability difficulties and will be hard to support far into the future. A new modern, more capable system is required with increased frequency coverage and improved bearing accuracy. The new ESM system is to have a potential for growth and the future addition of an ELINT capability.

Under ALEP, the Aurora's navigation and communications systems will be upgraded. A new navigation system, composed of twin, ring-laser gyro INS sets integrated with an embedded GPS sub-system, replaces the aircraft's current ASN-505 inertial navigation system and ARN-511 OMEGA and APN-510 Doppler navaids. A new communication system is planned and will provide a greater capability and built-in expansion. The new COM system integrates an advanced narrowband digital voice terminal (ANDUT) capability for secure HF SATCOM and selected new UHF radios. A new control and display system is planned to manage both the navigation and communication systems and will be located at the Aurora's pilot's, co-pilot's, TACNAV and NAVCOM positions with all-station listening capability through the intercommunications system.

One important improvement to the Aurora under ALEP is a new data management system to replace its existing obsolete general processing digital computer. The new processor is to be a COTS system with greater computing power, flexibility and built-in growth for future requirements. The new processor will be controlled through the installation of six new tactical workstations.

The new workstations are to be high-definition, color display, multi-functional sensor stations, allowing for any sensor operator to perform his duties at any console. The modernized tactical workstations will retain the unique "U" shape configuration that is indicative of the CP-140 Aurora.

One new improvement to be established under ALEP, is a new "Electro-Optical Suite." The new sensor suite replaces the existing OR-5008 forward looking infrared set with a new multi-sensor electro-optical system comprised of an improved FLIR, low light level television (LLLTV) and a new laser illuminator for an active-gated TV (AGTV) system. The EOS is expected to be an integrated system with the optical subsystems housed in a retractable ball-turret like the current infrared sensor.

Other Aurora life extension initiatives include addressing the fatigue life of the airframe and have led to the installation of Structural Data Recording Systems (SDRS) into the Aurora fleet to generate information for a structural life extension program study. A SLEP study would be important to determine the cost and effort required to prolong the structural fatigue life of the aircraft to the year 2025.

CP-140A ARCTURUS

Canada is one of a small number of Orion operators to have established variants of their Orion aircraft. In 1989, Canada

Arcturus in delivery forest-green primer. *Lockheed*

ordered three additional Orion airframes from Lockheed and eventually converted them into a specialized long range surveillance Aurora derivative designated CP-140A Arcturus.

Originally a requirement for six additional CP-140 Auroras was established under a Canadian defense policy document known as "Challenge and Commitment." This led to the Long Range Aircraft Project (LRAP) proposal to acquire six more CP-140s to augment the Aurora fleet due to the accumulation of high fatigue hours in the airframes. But the Cold War defused and the tempo of operations decreased. The need for fully equipped aircraft dissipated and the LRAP project was canceled. Later the Department of National Defense decided to purse an unsolicited proposal from Lockheed to purchase the last three Orion airframes off the Lockheed production line. These were to be stripped airframes without sensors and systems. The airframes were actually laid down into production for an anticipated order by Portugal back in 1985, but Portugal ended up selecting six ex-RAAF P-3B airframes converted to the Portuguese P-3P configuration. From that point onwards, Orion production continued to be three planes over that which were already sold.

The DND's intentions were to acquire the three airframes as needed pilot trainers to help reduce the wear and tear on the fully-equipped Auroras. The DND had previously considered acquiring the ex-Australian Bravos back in 1985-86 and converting them into pilot training aircraft, but the cost of purchasing the aircraft and converting the cockpits to the Aurora configuration was prohibitive at the time.

Later in 1988-89, the DND announced that the CP-121 Tracker (the Canadian version of the S-2) had been selected to be retired in the wake of military budget cuts. The Tracker had been the principle fisheries patrol aircraft and northern sovereignty patrol platform since the government mandate that the military to maintain a presence in the Canadian Arctic region back in 1985. The Auroras were to be assigned to the Trackers missions and this provided added justification to acquire three additional airframes.

Hence, as the acquisition process continued, the DND began promoting the three additional airframes as specialized variant aircraft of the Aurora to take over the Arctic northern patrol mission. It was at this point that the project to acquire the three airframes became known as AMSA, or the Arctic Maritime Surveillance Aircraft project. Later under AMSA, the three aircraft would be designated CP-140A and dubbed Arcturus.

The Arcturus configuration, devoid of all of the Aurora's ASW systems and sensors, is comprised of an APS-507 maritime surveillance radar (the Canadian version of the APS-134 non-imaging ISAR radar), an navigation suite consisting of ASN-505 inertials and ARN-511 OMEGA , a ARN-118 APN-510 Doppler navaids with a communication system including ARC-156 UHF, ARC-197 VHF and ARC-153 HF radios. Although the Arcturus were promoted as dedicated Arctic surveillance platforms, their configuration was capable of little else than that of pilot training flights.

The Arcturus themselves are based on the P-3C Update III airframe and retain the aircraft's 48 external sonobuoy chutes (non-functioning), as well as the Charlie's internal sonobuoy CAD and free-fall sonobuoy tubes and storage racks. The interior of the aircraft is modified similar to the Aurora with the "U shape" cluster tactical compartment, but are devoid of sensors and systems, except for the NAVCOM systems and the radar console. Each of the aircraft has been wired like the Aurora, as a provision for the possibility of modifying them into fully ASW equipped Auroras later if need be.

Delivered "green" out of the Lockheed production facility at Palmdale, California, the Arcturus were empty airframes, minimal equipment for the ferry flights to Industrial Marine Projects (IMP) in Halifax, Nova Scotia. IMP was sub-contracted to manufacture and install the aircraft's sub-flooring, flooring , racks, consoles, ceiling structures, interior trim and install the aircraft's lavatory, closets and galley-dinette area. The aft tail section of the Arcturus (past the galley) is equipped with tie-downs facilitating logistical transport of spare parts and equipment to deployed squadrons. Lacking the Aurora's ASW and associated gear, the Arcturus is lighter and profits from greater range and endurance.

With the delivery of the first finished Arcturus aircraft to CFB Greenwood on November 30, 1992 and the subsequent remaining two aircraft in April 1993, a single CP-140 Aurora was transferred to #407 squadron at CFB Comox, increasing its contingent of Auroras to five.

Near the end of the AMSA project, the government became aware that the new aircraft were not fully capable of performing their intended mission of conducting sovereignty patrols in Canada's Arctic. The DND explained that the Auroras were capable of conducting the mission and that the Arcturus provided valuable service for pilot training, but the government mandated that the Arcturus aircraft be capable of mission taskings. This set the AMSA office to find a means to provide a mission capability to the aircraft and led to the recent provisioning of the Arcturus with an interim data management system.

This is a laptop computer system based on an AST 900N notebook (340MB) computer with a serial port interface to a trimpack GPS unit. The interim data management system provides for manual entry of mission data during the course of a flight. Each entry is time coded as is the aircrafts position and track information. Programs within the unit include customized geographical map overlays of the patrol area, as well as undersea topographical contours. The system can also be used to generate post flight reports. The interim data management system facilitates utilizing the aircraft for fisheries patrols and counter narcotics missions besides pilot training.

As the future looms on the horizon, there are plans to upgrade the Arcturus, finally providing the aircraft with the capabilities to carry out missions they were intended for. The Arcturus upgrade program is actually an extension of the ALEP and includes the same improvements scheduled for the Aurora. The upgrade incorporates all of the new systems planned for the Aurora, except for the ASW suite. The Arcturus will receive the Spotlight SAR radar, new ESM system, NAVCOM systems, new Electro-optical suite and advanced data management system, as well as the new multi-functional tactical workstations.

The uniqueness of the Aurora/Arcturus in the Orion family of aircraft has proven itself and is now coming full circle. The "U" shaped cluster of the tactical suite, developed by a joint Lockheed-Canadian team, has proven to be the best tactical layout for an MPA aircraft. So much so that it is now being copied by other Orion operators and is the configuration selected for the next-generation Orion aircraft. Under the Royal Australian Air Force P-3 Refurbishment program (AP-3C), a variant of the Aurora's layout is planned. A similar Aurora type tactical suite layout was offered by Lockheed Martin in its entry for the United Kingdom's Replacement Maritime Patrol Aircraft (RMPA) competition to replace the long - lived RAF Nimrod. This entry, dubbed Orion 2000, by Lockheed Martin is also the proposed next-generation aircraft, replacing the Orion P-3C in the future. Possible operators include the US Navy.

CP-140A Arcturus on Patrol. ***CF/DND via VPI***

Portuguese Air Force P-3P Orion. ***PoAF***

PORTUGAL

Portugal is strategic in its location. This area of the Eastern Atlantic was NATO's southern flank and both Portugal and Spain shared the maritime burden during the Cold War. Portugal and Spain flew overlapping air patrols to monitor the vast open ocean and the natural choke point between the Atlantic and the Mediterranean at Gibraltar—guarding against increased Soviet expansionism in western Africa.

The Portuguese coastline runs for over 600 miles and knows no limits to its development. The sea is Portugal's livelihood and its sea lanes must be protected. Today, the Portuguese Orions' area of operation runs in a maritime triangle, extending out from the coastal mainland 800 kilometers west to the forest-covered, black (lava) pebbled beached archipelagos of Madeira and the isolated, mist-shrouded island group of the Azores, back to the home shores of the Iberian peninsula.

Portugal, like many of its neighbors in the region, operates the P-3 for maritime patrol. The Portuguese Air Force Orions are designated P-3P (the P for Portugal) and are a unique hybrid of the P-3. The aircraft is a standard P-3 heavy weight Bravo airframe specially modified with a digitally enhanced TACNAVMOD configuration equipped with state-of-the-art sensors and avionics -some common to the P-3C Update II.

The P-3P incorporates an improved APS-134 non-imaging ISAR radar, an advanced AQA-7(v)9 acoustic system, dual LTN-72 inertial navigation units and a APN-124 digital (TACCO) display with on screen touch controls. Other avionics include a specially modified ALR-66(v)3 ESM system and an AYK-14 digital computer. with a 1553B data bus for increased capacity to the tactical displays and Data Link. The aircraft also has provisions for the AAS-36 IRDS infrared system and the AGM-84 Harpoon missile—now permanently installed. The aircraft also utilizes a mixture of P-3 Bravo and Charlie communication systems and controls, as well as communications gear common to Portugal located in the radio operators station forward on the port side.

The P-3P has several features that make it somewhat unique in the Orion family of variants. Besides its three-tone gray "orca" tactical paint scheme, the aircraft's ESM system incorporates a specialized spinning directional finding (DF) antenna housed in a ventral radome just aft of the wings. The P-3P also comprises a dedicated ESM operator. The position was incorporated into the aircraft as an afterthought to the design plan and splits the non-acoustic sensor station between two operators. One is solely responsible for the ESM system and the IRDS detection set, leaving the other operator to radar and MAD systems management. The tactical compartment retains its P-3 Bravo side-by-side bench workstation layout, comprising (aft to forward) the acoustic sensor operator #2, acoustic sensor operator #1, the TACCO, navigator and non-acoustic operator #1 (radar and MAD) and #2 (ESM and IRDS). The P-3P is also uniquely equipped with C-130 webbed seating, installed in the area of the starboard side forward observer station. This auxiliary seating facilitates transport of ground maintenance personnel to detachment sites.

The P-3P program began in July of 1985 with an order placed to Lockheed for six P-3 Orion aircraft by the Portuguese Air Force. Originally Portugal was interested in three P-3C Update I/II Orions, but ended up accepting six ex-RAAF Bravos converted by Lockheed especially for Portugal with avionics common to the P-3C Update II.

Portugal had operated twelve ex-Royal Netherlands Navy P2V-5 Neptunes for ASW up until 1977 when they retired the

P-3P Orion exterior features. ***PoAF***

last four aircraft. Although maritime surface surveillance and search and rescue missions were conducted by various other Portuguese aircraft including F-27, CASA-12 and C-130s, there was no ASW capability. The inability to uphold their part of the NATO commitment prompted the initiation of the P-3 acquisition and the re-institution of ASW after a thirteen year hiatus.

The six airframes to be modified were among ten ex-Royal Australian Air Force P-3 Bravos that were traded back to Lockheed towards the purchase of ten new production P-3C Update II Orions in 1981.

Originally, eight of the ten RAAF Orions were scheduled to be sold to Argentina. The formal signing of the Lockheed contract with the Latin American country was scheduled for April 4, 1982. History now records that several days before (April 2nd) Argentina invaded the Falkland Islands. Australia, allying herself with the United Kingdom as a common-wealth nation (and having possession of the Bravo Orion aircraft) refused to permit any transfer of the planes to the Argentineans. The ex-RAAF Bravos remained in Australia until 1985 when a Lockheed proposal offered the aircraft to Portugal, whom at the time was evaluating a potential purchase of three P-3C Orions.

In October of 1985, the first two ex-RAAF Bravos were ferried from South Australia to Portugal and stored at the Portuguese Air Base at Beja. The next month, two more aircraft were flown out of Australia with another one in December. Unlike the first aircraft, the last one was bound for California and the P-3 Mod shop at Lockheed's Palmdale production facility. The program was under a direct sales contract with Portugal, including some US Navy FMS logistical agreements. Lockheed designed and re-engineered the aircraft into a prototype P-3P. This prototype was subsequently flown to the POAF airbase at Montijo, Portugal, in August 1988 and was used as the basis for the modification of five additional ex-RAAF P-3B. The conversion of the five Bravo Orions as performed by Oficinas Gerais de Materiel Aeronautical or OGMA, the Portuguese government aircraft rework facility. It was at this point that a delay ensued and it was another year before the first aircraft could be inducted

Close-up of P-3P ESM radome. ***Willem de Vreugd via Lockheed***

P-3P prototype aircraft. ***Lockheed***

P-3P current "ORCA" paint scheme. ***Marco Borst***

into the OGMA Mod shop. This delay provided more time for OGMA to set up and prepare its conversion line and also gave the aircrew training/transition program more time with the prototype aircraft.

The training program was established in two parts. The first phase began back in November of 1986 in Portugal and included utilizing the unmodified P-3B aircraft for pilot training, crew familiarization and aircraft safety flights. Phase two got underway by mid 1988 and consisted of ground crew and mission training. Along with time spent in the classroom, aircrew students used the P-3P prototype as a static Ops trainer while the ground crew classes utilized it as a maintenance training aid. Some of the acoustics and ESM students were sent to NAS Valkenburg in the Netherlands for systems training courses especially adapted for the P-3P.

Since their introduction of the P-3P into operation, the Portuguese have more than made up for all the years spent without an ASW capability. The Portuguese crews have proven their skills well in many annual NATO exercises, often outperforming a number of the other Orion equipped participants. In more recent years, the Portuguese Orions have participated in the United Nations' imposed sanctions against the former republics of Yugoslavia, as part of their NATO commitment. The commitment consists of one aircraft, sixteen crewmen and eight ground maintenance personnel deployed every two weeks on a rotating basis. The Portuguese contingents have maintained a solid ninety-five percent operational readiness status and were praised by UN and NATO controlling authorities.

As with other nations, Portugal's P-3Ps are finding new roles to perform in the wake of the Cold War. These missions include national interest type taskings such as coastal patrol, EEZ patrols, fisheries and pollution control as well as drug interdiction flights.

To effectively meet these and future military operational requirements, a number of low-cost systems improvements are being planned. One plan upgrades the aircraft's radar from the APS-134 to an imaging APS-137 ISAR radar, while another updates the existing ALR-66(v)3 ESM system to the ALR-66 B(v)3 system, increasing the system's targeting capability and improving the threat library. Another improvement for the P-3P encompasses adding a GPS system, a new MAD integrated with self-compensating capabilities for longer range localization and higher sensitivity parameters.

As maritime patrol gives way to sea control and missions operating close to hostile coastal territories, subject to threats from land-based interceptors, ground-to-air missiles defenses and small, fast patrol boats, improved weapons and self-protection systems are needed to counter these threats. Portugal recognizes these requirements and is pursuing low cost alternatives to the Harpoon, as well as counter-measures equipment. Proposed self-defense systems for the P-3P include chaff-flare dispensers and a infrared missile warning system. Separately, the Portuguese are also considering the acquisition of electronic intelligence equipment, giving the P-3P an additional ELINT capability.

JAPAN

Of all the international countries in the P-3 Orion fraternity, no one is more dependent on the sea than Japan. Ninety percent of its energy and industrial materials and seventy percent of its food are imported or come from the sea. The P-3 Orion provides Japan the means to protect its sea lanes of commerce, lines of communication and defend its home Islands and surrounding territories.

For many years Japan's maritime self defense force (JMSDF) searched for a follow-on ASW platform to replace its fleet of patrol aircraft made up of S2F-1s and P2V-7s, as well as their domestically produced turbo-propped P-2J Neptune. The JMSDF was concerned that their ASW capacity was deteriorating in the wake of new Soviet submarine advances. A study was initiated in the early 1970s to evaluate various current foreign and domestic ASW aircraft and a combination of hybrid proposals in order to maintain the JMSDF's ASW capabilities far into the future. The Japan Defense Agency looked at different foreign ASW aircraft, including the P-3C Orion, S-3A Viking, MKII Nimrod, proposed Atlantic IIB and the CP-140 Aurora, which was still under development at the time. They evaluated modifying existing domestic P-2J, PS-1, C-1 and Boeing 737 aircraft, as well as developing a completely new domestic aircraft and systems. They even considered foreign hybrids comprised of the P-3C or S-3A/CP-140 electronics installed onboard a domestically-developed aircraft.

A panel of experts studied the different aircraft proposals for adaptability to current forces and cost effectiveness. A second round of evaluations compared the P-3 Orion, the CP-140 Aurora and a combination of domestic-foreign hybrids. The CP-140 Aurora was still under development and would not be available for speedy integration into the fleet. The domestic-foreign hybrids proposals were not deemed cost effective and were eliminated from the study. Concerns over the rapid improvements in Soviet submarine technology mandated a new ASW platform be deployed as soon as possible

The first Japanese P-3C Orion produced by Lockheed. ***Lockheed***

for the JMSDF to overcome the threat. Hence the P-3 Orion was chosen. The Japan Defense Agency announced the acceptance of the P-3 Orion Update II.5 Orion on December 29, 1977.

Due to the sheer number of aircraft required by the JMSDF, approximately 100 aircraft, an agreement with the US Government and Lockheed arranged for the majority of the aircraft to be manufactured in Japan under license from Lockheed. The first three aircraft were produced by Lockheed. The next five were "Knock Down" airframes that were produced by Lockheed and shipped to Japan for assembly by Kawasaki Heavy Industries. The three complete P-3 Orions

Japanese Maritime Self Defense Force P-3C Update II.5 Orion accompanying a Japanese P-2J Neptune, which it replaces, over Mt. Fuji. ***JMSDF***

Lockheed produced "knock down" P-3C Orions for Japan on the production floor. ***Lockheed***

Kawasaki H.I. / JMSDF EP-3. ***JMSDF***

were delivered to the JMSDF during May 1981 with the five Knock Downs transported to Japan by the end of 1982. Then, under the license agreement, Kawasaki began full scale production of P-3 Update II.5 Orions by the middle of 1983. By early February 1991, Kawasaki began incorporating the Update III avionics upgrade into their production aircraft.

In early 1992, the JMSDF's Maritime Staff Office began seeking approval for a future Upgrade Program to improve its current P-3C Update III Orion. Some of the equipment proposed for the Japanese Orions included adding an improved ESM system and domestically-produced GPS and SATCOM systems, as well as a new data processor. The MSO selected the ASQ-212(J) computer group, which includes a Unysis CP-2044 data processor built in Japan under License. An indigenous SATCOM system has also been incorporated into Kawasaki production since 1993 with the CP-2044, ESM system and GPS unit introduced during 1996. These systems improvements are to be further retrofitted back into the existing sixty-eight P-3C Update II.5 Orions as they continue to be upgraded by Kawasaki to the Update III configuration during depot level maintenance.

A total of 101 ASW configured P-3C Update II.5 and Update III Orions have been produced by Kawasaki, with one hundred originally ordered by the JMSDF plus one as a replacement aircraft for a P-3C Update II.5 Orion (# 5032), lost in a wheels up crash on march 21, 1992.

In other developments, defense budget cuts have also been plaguing the JMSDF. In 1996, the JMSDF announced

JMSDF P-3 Orions ; first four from VP-3, next three from VP-6 and the last aircraft from VX-51. ***JMSDF via VPI***

JMSDF P-3C Orion over northern waters. ***JMSDF***

that there would be force reductions in the Japanese operational Orion fleet. Upwards of twenty aircraft are to be taken out of service and mothballed. At this time it is unclear as to whether it will be twenty of the newest Update III aircraft, twenty of the oldest Update II.5 or aircraft yet to be converted to the Update III configuration. The planned reduction may also include the disestablishment of a whole squadron or the equivalent through the reduction of one crew from each squadron. Currently there are nine maritime patrol, two training and one development squadron(s) flying Japanese P-3 Orions under the command of the JMSDF. They operate from five air wings based at locations throughout the home islands and conduct missions out into the East China Sea, Sea of Japan and Hokkaido area.

Kawasaki has also produced a number of different P-3 Orion variants for the JMSDF. The first variant was introduced in 1990 with the delivery of the initial EP-3 Orion. The Japanese EP-3 is an electronic support measures aircraft based on a new Kawasaki production airframe. Its mission is to conduct passive collection flights, recording electronic signals intelligence. The JMSDF EP-3s have capabilities that are somewhat similar to that of U.S. Navy EP-3E ELINT Orions, but are visually different in appearance to those of the US Navy aircraft. Upwards of five JMSDF EP-3s have been produced and are flown by #81 Squadron—more are scheduled to be built.

In 1995, Kawasaki delivered its second P-3 variant to the JMSDF. This single aircraft was designated UP-3C Orion. The UP-3C is an airborne systems flying testbed aircraft based on a new production utility airframe developed for the test and evaluation of ASW equipment as well as the flight performance testing of avionics, airborne electronic systems and weapons. The new P-3 variant replaces an aging P-2J which had been providing airborne test and evaluation for more than 20 years.

Kawasaki H.I. / JMSDF UP-3C. ***JMSDF***

JMSDF proposed UP-3E special reconnaissance Orion based on older ASW P-3C airframes. ***JMSDF***

Pakistan P-3C Update II.75 Orion. ***Lockheed***

The UP-3C began as a plain airframe, devoid of ASW equipment and sensor systems. The aircraft is equipped with a data management and display system, data collection and test instrumentation equipment, as well as radar, navigation and communications gear common to the ASW configured JMSDF Orions. The air test workstation compartment of the UP-3C is based on a US Navy P-3A/B side-by-side bench configuration consisting of (aft to forward) a NAVCOM operator, a radar operator, two test equipment operators, a test coordinator and lastly, an instrumentation operator. Other components include several instrumentation racks for test equipment with power and cooling outlets. The UP-3C was officially delivered to the JMSDF in August of 1995 and is operated by VX-51 air test and development squadron.

Another Kawasaki developed variant is the UP-3D Orion. The UP-3D is a new production utility airframe designed to provide electronic warfare training to Japanese surface fleet elements. This EW support (training) mission is similar to that of the American EP-3J Orion's command, control and communications-intelligence/counter-counter measures (C3I/CM) fleet training mission. Two UP-3D aircraft have been produced.

In 1995, the JMSDF announced the development of a forth Orion variant, the UP-3E. This is a proposed surveillance/reconnaissance version of the JMSDF's ASW P-3 Orion with a mission that is closely related to the US Navy's Special Projects Orions. Possible reconnaissance/surveillance systems to be employed include SLAR, SAR and ISAR radars, as well as ESM, EOS and laser systems. The new UP-3E Orions will be based on old production ASW P-3C airframes (presumably taken from those P-3C airframes being retired) and modified by Kawasaki for their new mission.

All totaled, the JMSDF's Orion fleet numbers one hundred and nine Orion aircraft—the largest foreign P-3 fleet in the world.

PAKISTAN

In July 1988, as Norway was preparing to receive the first of four P-3C Update IIIs, an order was placed with Lockheed for another three P-3C Orions. This request came from the middle-eastern nation of Pakistan. Pakistan was the first pro-western, predominantly Moslem, middle eastern country since Iran to request and receive P-3 Orions. Strategically located next to Iran, Pakistan also shares borders with Afghanistan and Kashmir to the north, India to the east with the Arabian Sea to the south. Pakistan has over five hundred miles of coastline along the Arabian Sea with direct access to the Indian Ocean.

Pakistan P-3C on VP-30 (NAS Jacksonville) flight line. ***US Navy***

Pakistan P-3C ***Lockheed***

Although built among the last P-3C Update IIIs for the US Navy (and the four Update IIIs for Norway), the Pakistan Orions are configured along the lines of P-3C Update II.5, except for its ARC-187 UHF and ARC-182 VHF radios, navaids and the ALR-66(v)2 ESM system that are more characteristic to the Update III. The Pakistan P-3s are also equipped with the APS-134 non-imaging ISAR radar and only have provisions (are wired) for the Harpoon. With the incorporation of these differences, the Pakistan aircraft have come to receive the unofficial designation of P-3C Update II.75 for possessing a configuration that is more than a Update II.5 but not quite an Update III.

The Orions were to be operated by #29 MPA squadron based at Mehran AFB. The P-3 aircraft were actually to be the custody of the Pakistan Air Force, despite having Navy markings. The air force was tasked to fly the aircraft for the Pakistan Navy, which encompassed air force flight deck crews in the cockpit and navy personnel manning the tactical compartment.

With the Pakistani Orion production well under way, P-3 transition training began at VP-30 (NAS Jacksonville) in January of 1991. The First aircraft was delivered out of the Lockheed production plant in November 1990 and was sent to NAS Jacksonville. The other two aircraft quickly followed and were utilized for pilot training at VP-30.

But during 1990, the US Congress had been examining several reports that Pakistan, Iraq and Iran, as well as other nations, were developing nuclear programs to produce weapons grade uranium. Congressional legislation was introduced to block the US government from selling any military equipment to these countries. The legislation, known as the Pressler Amendment (later the Pressler Sanctions) was ratified and passed into law by the US Congress during October 1990. It stated that before any military sales to the identified countries could take place, it must be accompanied by a certification from the (current) administration that the identified country does not posses nuclear explosive devices or the developing of such a capability.

For Pakistan, no such certification by the then Bush administration came forth and delivery of the Pakistan Orions was put on hold. By January 1992, shortly after the Orions arrived in Jacksonville and integrated into the training program, the aircraft were ordered into AMARC storage at Davis-Mothan AFB in Tucson, Arizona. On January 22, 1992, VP-30 crews flew the Pakistan Orions to AMARC where they would remain for long term storage pending resolution of the Pressler Sanctions.

In the meantime, the Pakistan Prime Minster Benazir Bhutto renewed demands for delivery of her Orions and other military equipment frozen by the Pressler sanctions. This included over one billion dollars' worth of F-16 Jets, Cobra helicopters, air-to-surface and surface-to-surface missiles, as well as spare parts for the P-3. In April 1995, during a US visit with President Clinton, the administration promised Mrs. Bhutto to resolve the situation and initiated legislation to relax the sanctions against Pakistan. Several options were suggested to have Congress lift the ban or relax them for Pakistan. The administration's viewpoint was that lifting the ban on military aid would induce the Pakistani Government to halt development of a nuclear weapons program and achieve the goals of non-proliferation.

Senator Larry Pessler, author of the Pressler amendment, even offered a third party sales of the Pakistan military equipment. Nations considered as possible buyers for the Pakistan P-3s were the Philippines, Singapore and Taiwan.

By the fall of 1995, legislation was introduced into Congress (for the administration) to permit the transfer of the equipment to Pakistan. After much debate the amendment passed the senate in September 1995 and then by the full Congress in a bill passed during January 1996. As of this date the administration has the full authority to proceed with the transfer of equipment to Pakistan. In December 1996, preparations were being made to fly the Orions to Pakistan in early 1997.

CHILE

Since the late 1980s, the US Navy has been receiving inquiries for surplus P-3 Orions from militaries all over the world.

Chilean UP-3A. ***Bob Shane***

One of the first was Chile. Chile was in need of a new maritime platform for coastal patrol, search and rescue—coast guard duties, fisheries patrols and its new counter narcotics mission, as well as civilian taskings such as timber and natural resources observations.

Chile had been familiar with the P-3 for years through UNITAS, the annual joint US Navy/Latin American naval exercises, where the Orion's capabilities had been demonstrated to the navies of South America. In 1988, the Chilean Navy (Armada de Chile) initiated a replacement patrol aircraft program called OLYMPIA to upgrade and augment their fleet of Embraer EMB-111ANs. Besides considering the P-3 Orion, the Olympia program examined the possibility of utilizing surplus British Nimrod airframes.

After several meetings and site surveys of P-3As in desert storage at AMARC, the Chilean Navy made an official request to the US Navy for eight to twelve P-3A Orions in 1992. By the next year the Chilean Navy request was approved for eight UP-3A Orions. The Chilean UP-3As were selected from a specific number of straight (non-updated) P-3A and UP-3A Orions stored at AMARC. The aircraft were stripped of ASW equipment and the de-lethalized weapon control systems.

Under the ensuing FMS case, the Chilean Navy opted for a civilian contractor (TRACOR, Inc.) to train Chilean crews and provide logistics. The Orion's airframes were taken from AMARC, at Davis-Monthan AFB in Tucson, Arizona, to the nearby re-work facility at Western International Aviation Inc. Western International was contracted by the Chilean Navy to rejuvenate the aircraft from storage and performed airframe inspections, repair and painting prior to ferrying the Orions to their new home in Chile—at the naval air base at Vina del Mar, located 60 km outside of Santiago.

The Chilean Navy took delivery of the first of eight UP-3As on March 3, 1993. The first Chilean P-3s arrived just in time to participate in the Chilean 70th Naval Anniversary celebration and represented the remaining seven yet to be delivered. The last aircraft was accepted by the end of 1994.

Of the eight aircraft, upwards of six are believed to be operational with the remainder scheduled to be utilized for maintenance training and parts aircraft. The aircraft are expected to be equipped with Chilean avionics, including radar and communications systems and a proposed COMINT suite.

Interior configurations are to vary between aircraft, with at least one operational aircraft to be utilized for a utility transport role. There has been some suggestion that several of the operational aircraft will be equipped with an Israeli developed weapons systems—a capability not originally permitted by the US Navy.

THAILAND

The Royal Thai Navy's Orion program began in 1988 with the offer of P-3 Orions to upgrade the Nation's fleet of maritime patrol aircraft, but a period of instability arose within the Thai government and caused uncertainty towards the advancement of the program. After a delay of several years, and re-

Royal Thai Navy P-3T Orion. ***P-3 Publications***

P-3T Orion. *P-3 Publications*

assurance of stability within the government, the Thai P-3 program re-emerged in the fall of 1993 with the signing of a US Navy FMS case. The signing initiated the delivery of two P-3A aircraft to the Royal Thai Navy base at Utapao, Thailand, during the first week of January 1994. The RTN Orion aircraft modification program commenced later that same month with the induction of a P-3A into the Mod shop of the Naval Aviation Depot at NAS Jacksonville, Florida.

Since the end of the Cold War, the Royal Thai Navy has been evolving from a coastal navy to a blue water one. The RTN ordered a number of frigates from China in 1988 which prompted a parallel interest in the P-3 Orion to fulfill its offshore operations and anti-surface warfare (ASUW) mission requirements. It's this interest in the P-3 that has spawned the recent acquisition of the Orion aircraft for the Royal Thai Navy. Thailand, like so many other nations, recognizes the P-3 as a means to expand and enhance their maritime patrol capabilities. The plane's multi-mission capabilities offer a certain degree of insurance against possible future regional instabilities.

The Thai Orion Mod program at NADEP Jax consisted of performing special rework and depot level maintenance on the airframes to bring three Orion aircraft up to full mission capable status. The first two aircraft, re-designated P-3T (the T for Thailand) are basically TACNAVMOD ASW configured P-3As with a number of additional equipment and system upgrades. These improvements include a new LTN-72 internal navigation set, new navaids (ARN-118 TACAN and LTN-211 OMEGA), a new ESSI radar monitor system display for the existing APS-80B radar and provisions for IRDS and the AWG-19 HACLCS Harpoon control system.

The RTN is planning to equip their P-3Ts with an infrared sensor and have been interested in the FLIR Systems, Inc. AAQ-22 SAFIRE thermal imaging system. As to the P-3Ts being wired for Harpoon, this is based on the RTN already possessing Harpoon control systems and missiles in their existing patrol aircraft.

Although the P-3Ts were delivered without an acoustic system capability, a number of AQA-7(v) 4/5 systems were sent crated to Thailand for subsequent domestic installation, once a training program could be established.

The third Orion aircraft, re-designated UP-3T, underwent special conversion rework to convert the aircraft into a utility transport version of the P-3T. This Mod consisted of stripping the airframe of all sensor stations, ASW equipment and associated sonobuoy system gear. The aircraft's floor was then structurally strengthened and modified with floor tracks to accommodate passenger seating and tie down attachments for cargo transport. The tracks run aft from the forward left hand equipment rack back past the main cabin door. The aircraft also received an all new interior refit with Gray wall coverings and trim.

The UP-3T configuration is somewhat unique in the P-3 fraternity in that it retains a limited surveillance mission capability through the addition of a SENTAC station. The SENTAC station, developed by NADEP Jax for Thailand, incorporates elements from both the P-3T's sensor #3 and TACCO stations. The station includes the ESSI radar monitor display and controls for the APS-80B radar, TACCO interval computer, and navigation and navaids, as well as control provisions for IRDS. The SENTAC station is located forward on the port side in the old radio operator's station. Thailand has plans to eventually refurbish the aircraft in-country for VIP transport duties. Additionally, all three aircraft will be equipped with the PRIMUS 400 commercial color weather radar.

The two remaining baseline P-3As, previously flown to Thailand at the start of the Thai Orion program, are to be utilized as ground maintenance trainers and as future spare parts aircraft.

P-3T Orion. *P-3 Publications*

Korean P-3C Update III. *Lockheed* via VPI

Greek P-3B ***Lockheed***

Operationally, the Thai P-3s fall under the command of the Royal Thai Navy's Air Wing One located at the RTN air base, Utapao. The Wing is composed of four maritime patrol squadrons; #101, #102, #103 and #104, operating a mixture of Dornier 228, Fokker F-27 and Grumman S-2F Tracker patrol aircraft. Mission taskings include coastal patrol and surface surveillance missions with capabilities for anti-surface warfare (ASUW) and ASW.

Squadron #101 has been selected to become the operating unit for the P-3T. Over fifty RTN aircrew and ground maintenance personnel were sent for training to VP-30 at NAS Jacksonville during the fall of 1994.

The first P-3T was turned over to the Royal Thai Navy during a formal ceremony at NADEP Jacksonville on February 6, 1995. The second quickly followed and both aircraft were flown to Thailand ten days later to begin operations with #101 Squadron. The UP-3T was delivered during the fall of 1995.

REPUBLIC OF KOREA

On December 12, 1994, the first P-3C Update III Orion for the Republic of Korea Navy made its maiden flight. This event also marked the flight of the first Orion aircraft off Lockheed's Marietta (Georgia) production line, the third Lockheed production assembly line to produce the P-3 Orion since its introduction to the US Navy in 1962 (the first being Burbank, California, and the second at Palmdale, California).

The ROKN Orion program began four years earlier in December 1990 with the announcement that the Republic of Korea would be ordering eight P-3C Update III aircraft, thus prompting the re-opening of the P-3 production line at Marietta. Lockheed had been undergoing re-structuring at the time and generated the subsequent move from Palmdale to Marietta.

The P-3C Update IIIs that the Koreans received were not the same production Update III aircraft previously produced for the US Navy and Norway in 1989. These Orions differed in that they were equipped with some of the new systems currently being retrofitted into US Navy Update III, such as the ASQ-212 central processing system (including the CP-2044 computer) and ALR-66A(v)3 ESM system. The Korean Orions are also equipped with non-standard Update III systems such as the APS-134 non-imaging ISAR radar and dual ARC-187(v) UHF radios. The ROKN P-3Cs also lack some of those systems common to the Orion for years, having only been provisions for the Harpoon and IRDS.

Besides the differences in mission equipment, the Korean version of the production P-3C incorporates more modern manufacturing techniques with emphasis on corrosion prevention and protection, as well as the utilization of alternative component materials.

The ROKN Orion mission is one of defense, guarding against hostile attack from potential adversaries in the region. With an operational area that runs from the western Sea of Japan through the eastern Yellow Sea, including the vital Tsushima straits choke point, ROKN Orions are tasked with surface and sub-surface surveillance to prevent against disruption of its all important sea lanes and potential amphibious assault of its coastal territories.

The Korean P-3C provided a leap in technology and new maritime patrol capabilities to the ROKN, over the older Grumman S-2 Trackers that they replaced. The ROKN Orions are operated by #613 MPA squadron under the command of Air Wing #61 based at the naval air station PoHang, South Korea.

The ROKN has a operational requirement for sixteen P-3 Orions. They have already secured funding for another eight

Greek P-3B Orion; note HU-16 Seaplane in the background, which the Orion replaces. *Logistics Services International*

with ongoing efforts to procure additional funds. It's not yet clear as to what P-3s will be available if and when Korea is ready to order new aircraft. Will it be more P-3C Update IIIs, built specifically for them, or the new proposed Orion 2000?

In the meantime, there has been talk of the ROKN acquiring US Navy surplus aircraft to be modified as pilot trainers. As of early 1997, no official request has been received by the US Navy for aircraft.

GREECE

Greece is among the newest members of the worldwide P-3 Orion community with the delivery of the first of six P-3Bs and four P-3As during the spring of 1996. This acquisition stems from a 1990 defense cooperation agreement between the Greek government and the United States that allows for the retention of the US Navy base at Souda Bay, Crete, and a number of existing communications sites throughout the country. In exchange for base rights, Greece receives twenty-eight F-4E Phantom jets, twenty-eight A-7 Corsairs, four "Charles F Adams" class guided missile destroyers and a number of P-3 Orions.

Originally, Greece was interested in six to fourteen P-3As to be operated by the Hellenic Air Force. Proposed utilization of the aircraft included troop transport and as a platform for parachute jump training. Other considerations encompassed converting some of the P-3s into airborne fire-fighting air tankers. It seems that Greece is plagued by seasonal forest fires in the summer each year, much like the western United States.

All totaled, eight aircraft were to have been operational missions with one configured for utility taskings. But the deal fell through due to, among other things, in fighting between the Hellenic Air Force and Hellenic Navy.

Then the Hellenic Navy expressed an interest in acquiring twelve P-3As for maritime patrol to replace their aging fleet of HU-16 Albatross amphibious seaplanes. Of the twelve aircraft proposed, six were to have been MPA capable with six airframes stripped for spare parts.

It was at this time that the US Navy, in an effort to balance out capabilities in the region, offered Turkey an equal number of P-3A Orions for maritime patrol.

Meanwhile, the Hellenic Navy's maritime capability was beginning to diminish with the grounding of several of its HU-16 seaplanes. This led to a 1992 US Navy approved plan to lease four TACNAVMOD P-3 Bravos to Greece. This FMS case was an interim measure to provide the Hellenic Navy with aircraft until they could take delivery of the P-3A from the ongoing FMS case.

The proposed Bravo lease was to have included a separate FMS logistical case to furnish training and logistics for the P-3B and provided for early delivery of P-3As from the Alpha FMS case as spare parts aircraft. The Bravo FMS case was signed on February 16, 1994 and incorporated the lease of four ex-USN reserve TACNAVMOD P-3Bs and the acquisition of two TACNAVMOD P-3As as ground trainers, as well as two straight Alphas to be utilized as spare parts aircraft.

By this point, the US Navy received word that the Turkish Navy would decline the Navy's 1991 offer of twelve P-3 Alphas. The capabilities of the P-3 exceed the MPA requirements of the Turkish Navy that had been flying the S-2E Tracker. Although no particular follow on aircraft was identified, the capabilities of the S-3 Viking are more in line with Turkey's requirements and several unofficial inquiries have been made to the US Navy as to S-3 availability.

The announcement made by Turkey caused the US and the Hellenic Navies to re-evaluate the Greek FMS case. A revised FMS program canceled the Bravo lease and provided the grant of four P-3Bs to Greece. Under the new FMS case, the four Bravos were to be operational maritime patrol aircraft with the four P-3 Alphas as ground trainers and spare parts aircraft. The new program also includes the option for two additional TACNAVMOD Bravos for operational use. This brought the total Greek Orion acquisition to six P-3 Bravos and four P-3 Alphas.

The Greek P-3s are based on ex-US Navy reserve (light weight) TACNAVMOD (block 1) P-3B Orions slightly enhanced with the AQA-7(v)11 acoustic system and the provision for IRDS infrared sensor. Standard avionics include APS-80 radar, ASQ-10 MAD and ALD-2 ESM (only one of the aircraft is fitted with an ALQ-66(v)2 ESM system) The Greek Orions are also equipped with -14 engines. These are actually T56-A-10WA turboprops converted via Allison's -14 conversion kit.

The Greek P-3 Mod program began in July 1995 with the induction of the first Hellenic Navy P-3s into the Mod shop at Chrysler Technologies Airborne Systems, Inc. Facility in Waco, Texas—now a division of Raytheon E-Systems. The Waco Facility was tasked, via a US Navy Omnibus contract providing augmentation support to the Navy's naval aviation depot (NADEP) facilities, with conducting phased depot maintenance (PDM) on the airframes and a complete overhaul of landing gears and engines. This includes inspections, rework and repair of the airframes with emphasis on corrosion control. The interiors were refitted with new wall coverings and trim, curtains, floor boards, equipment bay doors and galley fixtures as part of an overall habitability enhancement. Finally, the Waco Division painted the Greek Orions in standard USN all-gray tactical paint scheme with Hellenic Navy markings.

Completion of the first Greek Orion occurred on May 22, 1996 with the subsequent delivery of the aircraft to Greece by the end of the month. The remaining aircraft were all delivered by summer's end. Of the four Alpha aircraft delivered to Greece in early 1995, two will be utilized as ground maintenance trainers, while the others have been disassembled by Hellenic Aerospace at Tanagra, Greece, for spare parts.

The Greek Orions are to be operated by #353 Squadron which is part of 112 wing based at ELEFSIS Air Base located northwest of Athens. Like a number of smaller force Orion operators the Greek Orions are to be piloted by Hellenic Air Force personnel with the tactical compartment manned by Hellenic Navy sensor operators.

Close-up of Greek P-3 markings. *LSI*

Chapter 5
Orion the Versatile

As with other aircraft in the world, the P-3 has spawned many different variant models. Besides those gener ated by foreign operators, there are P-3 aircraft with specialized airframe and mission system configurations that vary from the standard production P-3 Orions.

The majority of the Orion variants encompass those specially modified for the US Navy with a small number specially developed for US Government agencies' missions and civilian contractor applications. The US Navy has been responsible for creating most of the variant configurations over the Orion's thirty (plus) year history, taking advantage of the Orion's inherent operating capabilities to perform specific new dedicated tasks.

The first variant of the P-3 was unique in that until quite recently no one knew of its existence. Although considerably modified, the aircraft was kept secret for over thirty years and only here is it described in some detail for the first time.

The BLACK ORIONS

Operational for less then two years, in 1963, three P-3A Orions were borrowed from the US Navy and were eventually uniquely modified by the CIA for "black flights"—initially over mainland China. Black flights are clandestine airborne activities in support of covert CIA operations that encompass various missions ranging from border periphery surveillance flights, electronic surveillance and cross-border photographic reconnaissance missions to psychological warfare leaflet drops and agent infiltration paradrops. Some black flights include the worldwide transport of arms, agents, special VIPs and operational support equipment circumventing international customs inspections.

In June 1963, a maritime patrol P-3A (#149673) was diverted from training flights at VP-30, NAS Jacksonville, Florida, and flown to the Naval Aviation Depot at Alameda, California. There, a one-of-a-kind cargo door modification was performed. The Mod included widening the aircraft's main cabin door, adding a mirror image door next to the existing one. Both doors swung inward and back out of the way, resulting in an opening approximately 53 inches wide. Unfortunately, there was a subsequent lack of additional structural support members and the tail section of the aircraft nearly twisted off during a test flight. It's interesting to note that despite being eventually repaired, which included a body strap for extra support, the aircraft retained a 4° twist in its tail the rest of its unusual career. It was at this time that the Orion became immersed into the black world of classified projects and along with two other P-3As (#149669 and #149678) disappeared into temporary obscurity.

Over the next year, the three aircraft went through a series of airframe modifications conducted by such companies

NAWC Pax River NP-3D Orion. ***US Navy / NAWC Pax River***

Black P-3 Orion; note weapons bay leaflet spreader, ram air (sampling) scoops, short propellers, bubble dome and Sidewinder missile rails as well as a number of antennas and pods. ***E-Systems***

as Lockheed and E-Systems of Greenville, Texas. Both #669 and #678 were also equipped with a revised mirror image cargo door modification performed by E-Systems.

The black modifications made to the Orions provided the aircraft with new multi-mission capabilities that consisted of state-of-the-art electronic systems, and advanced avionics that would later lead to the electrification of the Vietnam War after 1969. Mission systems included a side-looking airborne radar (SLAR) for border periphery surveillance, electronic multi-wave band communications intercept equipment for the intercept and recording of Chinese communications and a multi-spectrum infrared detector for passive surveillance. It's also believed that a newly developed acoustic eavesdropping device was tested aboard the black P-3. The unit was so sensitive that it could detect engine and machine-manufacturing noises at long range. Photo reconnaissance had long been the backbone of airborne surveillance and the Black Orions were apply equipped for slant-range or oblique photography. Airframe modifications included the cargo doors for facilitating paradrops of equipment, arms, ammunition and agents during penetration (cross-border) reconnaissance flights and a weapons bay mounted (motorized) leaflet spreader capable of dropping tens of thousands of propaganda leaflets prepared by the psychological warfare compartment (office) within the CIA. The leaflet drops would take place during border periphery/penetration flights where the leaflets would drift down over mainland cities. Leaflet drops were common occurrences during multi-mission black flights, as was air sampling.

Under the direction of the Atomic Energy Commission, tasked at the time with collecting worldwide air samples in an effort to monitor and track the spread of nuclear weapons development in other countries, the Black Orions were equipped with air sampling/collection apparatus. The air sampling gear was connected to ram air scoops installed on both sides of the aircraft aft of the cockpit. Samples collected could not only indicate that an atomic/nuclear blast had been detonated, but could reveal radioactive particles that could determine the type and power level output of the blast.

The Black Orions were also configured for self-preservation. Due to its operational missions being conducted in very hazardous environments, the aircraft's turboprop engines had heat dissipating (extended) exhaust shrouds and for a while, shortened propeller blades to decrease the prop noise from the aircraft. The aircraft were also painted "black," as black as the world they operated in. The aircraft also bore, at least for a while, National Chinese markings. The probable premise was if they ever got shot down the Chinese markings gave plausible deniability for the CIA and the wreckage would most likely be identified as an L-188 Electra that the National Chinese were known to have for regional airline service.

One of the more interesting self-preservation modifications to the Black Orions was the addition of sidewinder missiles for self-defense against Chinese Migs. The Mod included missile rails mounted on the Orions' wing weapons stations, an observation bubble installed in the top of the fuselage in the aft section of the plane and a fire control unit setup in the cockpit. In 1964-65, these three would have been the only Orions in the world to have been equipped with sidewinder missiles and one is believed to have shot down a Mig fighter during this period.

The Black Orions were among a number of specially configured aircraft utilized by the CIA for clandestine black flights since before the Korean War. The Black Orions were flown from secret airbases on the Island Nation of Taiwan, flying off into the night on flights that would take them over mainland China. The CIA was most interested in communist China, from its development as an industrial nation to its proliferation of nuclear weapons and ICBM missile technology.

Although the CIA's U-2 activities were exposed in 1960 and the agencie's worldwide reconnaissance operations were somewhat curtailed, overflights of the mainland China region continued in earnest with the Black Orions (for a time) taking up the lead, starting in 1964.

The P-3 flew low altitude missions skirting along the southern Chinese coastline with occasional penetrations of the border to ferret out enemy air defense radar systems to map there location. Overland flights were conducted to locate military installations and airfields, as well as to record military communications traffic and gather data as to the level of China's industrial complex. Other overland flights included dropping leaflets and sampling the air besides paradropping agents or equipment and arms to indigenous counter-insurgents. The CIA later tasked the Black Orions with additional surveillance missions overland in Burma and up to Tibet, to gather data on Chinese military operations against the civilian populace there.

With the turboprop engines providing fast dash speeds and a heavy payload carrying capacity, the P-3 was actually capable of performing all the CIA's multi-mission taskings on one flight—and often did.

A typical mission flight would have the Black Orions leave Taiwan at such a time to put the aircraft close off the coast of mainland China by dark. With night enveloping the country, the P-3 would skirt along the coast recording communications traffic, electronic signals and industrial acoustic data. Other aspects such as photography, infrared imagery and radar information could also be gathered as the aircraft penetrated the coastline and ventured inland across the southern region of the country. At this point, additional data collection could be continued as preparations were made for agent or equipment drops. Leaflet drops could also be conducted at this point. At some point the aircraft would then cross the border into Burma and later up into Tibet to collect intelligence data before making the trip all the way back towards Taiwan, landing soon after daybreak.

In September of 1966, one of the Black Orions was detached from its operations over China and was assigned to a joint Department of Defense/Defense Intelligence Agency operation via the US Air Force. The aircraft was provided to support an intelligence-collection program in Vietnam codenamed "BENT AXLE." The program involved gathering intelligence to locate Guerrilla encampments and NVA troop concentrations along the Ho Chi Minh Trail system of Laos and North Vietnam. The Black Orions operated out of Okinawa, Japan, at night and flew over Vietnam, conducting one of the first electronic intelligence gathering missions of the conflict. After 1968-69, the incorporation of electronic surveillance/reconnaissance in the Vietnam conflict commenced under an ever widening program called "IGLOO WHITE."

This aircraft was only used for a short time, flying directly from its mission over Vietnam to the United States and the Naval Air Station Alameda in January of 1967. The two other Black Orions followed several months later.

The Black Orions existed between May 1964 and April 1967. The multi-mission Black P-3 had been something of an interim aircraft for the CIA. Operating between an era of somewhat mission-dedicated airborne platforms, previously used and newer sophisticated aircraft to come on line—namely the SR-71 which made its first test flight in 1965 and became fully operational by early 1967. The SR-71 took over most of the black flight missions over China and Southeast Asia.

Within months of the Black Orions' arrival back in the United States, the aircraft were back at E-Systems undergoing modifications that would establish the next official P-3 variants for the US Navy, the EP-3 Orion.

EP-3 ORIONS (EP-3A)

By the mid 1960s, the war in Southeast Asia had escalated and had stimulated the need for electronic warfare systems to support operations and gather intelligence. The intelligence demands placed an urgency on the development of new data collection systems and to improve the existing array of electronic intelligence (ELINT) devices. The EP-3 was born out of this need.

In early 1967, Lockheed was awarded a contract to develop a new replacement for the US Navy's EC-121 Constellation ELINT intelligence gathering aircraft. Partly on the success of the Black Orions' performance over China, the P-3 was considered the best replacement platform for the program. With the limited number of surplus airframes available, the parked Black Orions were the most logical choice as candidates for EP-3 conversion. The Black P-3s' capabilities were very similar to those required by the proposed EP-3 ELINT mission. Most of the initial reconfiguration of the aircraft was already done.

One of the Black aircraft (#149673) was developed as a proof-of-concept test aircraft for ELINT systems development and modified with a large ventral dish radome in the bombay and dorsal and ventral canoe radomes. This first aircraft was designated EP-3A. Although at this point the EP-3A signified a development platform for the follow-on ELINT configura-

EP-3E Orion. ***US Navy***

Naval Weapons Lab "Daulgren Bullet" EP-3A Orion. ***Isham collection***

tion, the subsequent reality was that the EP-3A later came to represent a electronic systems test and development aircraft used to trial and evaluate electronic warfare (EW) systems. After initial ELINT systems testing for the EC-121 replacement program, the aircraft was assigned to the Pacific Missile Test Center, or PMTC, stationed a NAS Point, Mugu. There, one of #673's first projects involved flight testing the Electronic Emitter Location System, or EELS. EELS was an airborne electronic system that passively detected and located the sources of anti-aircraft radar emissions to direct strikes against them. The system became the forerunner of the modern surface-to-air missile (radar) warning system eventually installed onboard B-52 bombers and other strike aircraft operating in Southeast Asia.

Later in September of 1971, #673 was transferred to the Naval Weapons Laboratory (NWL) in Dahlgren, Virginia, for various weapons test programs. It was at this time that the aircraft picked up the nickname of the "Daulgren Bullet." The EP-3A was deconfigured during this time, losing its dorsal canoe radome. This is an important point, due to the fact that

VX-1 "EMPASS" bird. ***M. Wada via Isham collection***

VX-1 "EMPASS" EP-3A Orion *B. Thompson via Isham collection*

NRL EP-3A Orion. *Baldur Sveinsson*

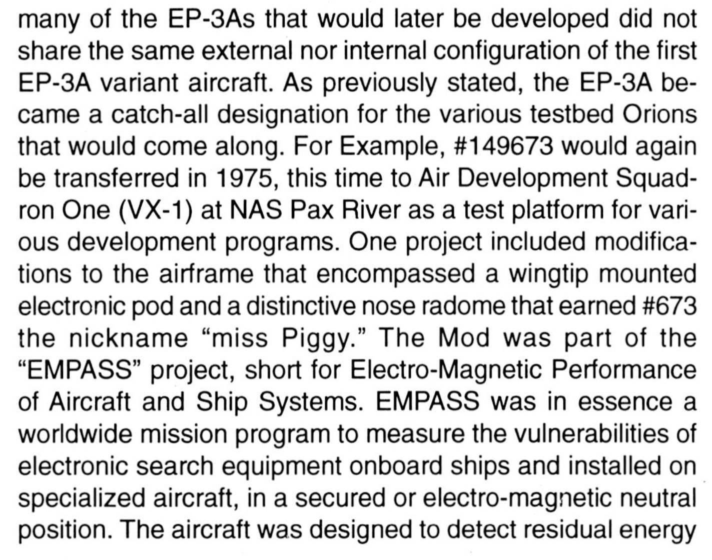

many of the EP-3As that would later be developed did not share the same external nor internal configuration of the first EP-3A variant aircraft. As previously stated, the EP-3A became a catch-all designation for the various testbed Orions that would come along. For Example, #149673 would again be transferred in 1975, this time to Air Development Squadron One (VX-1) at NAS Pax River as a test platform for various development programs. One project included modifications to the airframe that encompassed a wingtip mounted electronic pod and a distinctive nose radome that earned #673 the nickname "miss Piggy." The Mod was part of the "EMPASS" project, short for Electro-Magnetic Performance of Aircraft and Ship Systems. EMPASS was in essence a worldwide mission program to measure the vulnerabilities of electronic search equipment onboard ships and installed on specialized aircraft, in a secured or electro-magnetic neutral position. The aircraft was designed to detect residual energy leaks from search electronics in the off position that could give away the battle group's position.

There have been other EP-3As developed over the years with each airframe modified differently to a special tasking, test project or mission—sharing only the deletion of the MAD boom as a primary means to identify them.

One ASW P-3A to be recruited for duty as a PMTC test range aircraft was #149671. It was designated an EP-3A and served for many years with the Point Mugu test center until it reached the end of its useful service life and was retired. But P-3s are hard things to retire and have always found a means to survive. In this case, #671 was immediately acquired by the Naval Air Weapons Test Center at NAS China Lake, California (now NAWC-WD China Lake), and has been regularly utilized as a "static ground air flow generator." In other words, a wind machine with rushing air produced by the aircraft's turboprops to create the airflow necessary to test coefficient

Another NRL EP-3A. *US Navy / NRL*

VAQ-33 EP-3A Orion. ***M. Grove via Isham collection***

of drag or hydrodynamics over developmental missile systems. The airflow generating P-3 continues to perform to this day!

Most EP-3As were modified from maritime patrol P-3As given up by the fleet, although some have been developed from later P-3A variant airframes. This occurred with two Alpha variants modified specifically for other mission taskings that were subsequently acquired by the Naval Research Laboratory (NRL) and its Flight Support Detachment located at NAS Pax River, Maryland.

NRL has been the primary science and engineering research and development institution for the Navy since its establishment in 1923. Its mission is to conduct broad based, multi-discipline scientific research and advanced technological development that has direct applications toward Navy missions capabilities and requirements, and to probe physical science which will advance technology for the Navy.

NRL maintains the flight support detachment to provide the Navy scientists with the means to conduct worldwide extended airborne research and systems development. Of the aircraft acquired by NRL, one (#149674) was a weather reconnaissance version of the P-3 and was re-modified into an EP-3A utilized for the ASW Targeting and Bottom Contour Mapping Research Project. The aircraft would also serve as a platform for a variety of research programs that encompassed collecting oceanographic research data on water dynamics including salinity measurements, sea current mapping and temperature readings collection.

Another NRL EP-3A (#149670) was capable of collecting similar oceanographic research data, having been re-engineered from the first project Birdseye Orion variant flown by US Navy Oceanographic Development Squadron Eight (VXN-8) before making its way to NRL. Having been predominantly utilized for oceanographic research studies, both aircraft were equipped and operated for electronic warfare studies.

Each of the NRL Orions was a versatile testbed platform with common internal modifications encompassing floor & ceiling tracks for various electronic racks and additional power/cooling outlets throughout the cabin. Externally, the aircraft looked very different, with various shaped blister pods and radomes protruding along the fuselage that housed prototype radar systems, antennas and EW sensors.

Another EP-3A Orion (#150529) once operated by Electronic Warfare Aggressor Squadron 33 or VAQ-33 out of Key West, Florida. Looking nothing like its NRL brethren, this EP-3A was virtually devoid of all external blisters, canoes or large radomes save for sprouting an occasional EW pod attached

PMTC (now NAWC Point Mugu) EATS "Billboard" Orion. ***US Navy / PMTC***

Billboard Orions. ***C. Kaston via Isham collection***

to the aircraft's wing hard points. The mission of this EP-3A was EW countermeasures training to US Navy fleet battle groups. The aircraft simulated enemy maritime patrol and reconnaissance aircraft, as well as missile threats in a hostile EW environment.

Some of the most unique EP-3As were those developed by the Pacific Missile Test Center in 1979. Known as "EATS" Orions, the Extended Area Test System equipped EP-3As provided the capabilities to support PMTC's primary mission of testing missile systems. The EATS Orions extended the PMTC test range far out to sea, thus providing over-the-horizon tracking, targeting accuracy verification and operations data collection of missiles under development, such as the Trident and Peacekeeper missiles, as well as the Sidewinder, Sea-Sparrow and Harpoon tests. The PMTC EP-3As also provided a platform for other advanced airborne systems to be developed and tested.

Initially, three fleet ASW P-3A Orions (#150499, #150521 and #150522) were acquired and extensively modified. This consisted of removing existing ASW sensors, gear and systems before being outfitted with the usual test aircraft floor tracks, for roll-on electronic racks and the installation of power and cooling outlets. The EATS Mod then called for the fabrication of "Billboards," or horizontal extensions of the Orion's vertical stabilizers to house a Raytheon, Rotman-lens Phased Array. This modification was performed by Hayes International Corporation under the auspices of Tracor/FSI.

It's interesting to note that around the late 1970s and early 1980s, major advances in electronic technology emerged. Their impact on the EATS program was that more compact electronic systems could be built. Thus, all the PMTC range Orions could be equipped with the EATS system, without the need for the large Billboard extensions. Although the three initial Orions were modified with the Billboards, only the prototype EATS aircraft (#150499) was equipped with the Rotman-lens phased array. The other two Billboarded EP-3As (#150521 and #150522) were equipped with the new, more compact EATS systems, their horizontal extensions remaining hollow until the installation of a secondary telemetry system array several years later.

Two additional P-3As were also acquired by PMTC and re-configured as EP-3A range test aircraft. These Orions (#150512 and #150520) were further equipped with the Sonobuoy Missile Impact Location System, or "SMILS." The SMILS Orions' monitor passive sonobuoys laid in wide patterns on sea test ranges to accurately locate the splash-down,

PMTC "SMILS" EP-3A Orion. ***US Navy / PMTC***

PMTC "HARPOON" Orion. ***US Navy / PMTC***

impact points of test missiles. The system's data collected assisted in calculating the accuracy of the missile's targeting system and rate of trajectory.

Besides the two PMTC SMILS-equipped Orions on the west coast, VX-1 operates a SMILS-equipped P-3C Orion on the east coast to support missile tests and provide support to NASA Space Shuttle launches—locating the re-useable rocket boosters that fall back into the sea after liftoff and must be recovered.

Some of these same EP-3As were also equipped with the onboard Telemetry "Miss Distance" Indicator system that received, recorded and relayed missile test data back to ground controlling stations via secure UHF radios and now SATCOM. Most of the PMTC Orions were also provisioned with various optical photographic systems to record re-entry parameters or streak photography of the missiles being tested. Some of the photographic systems include forward-looking/ side-looking 70mm sequential, 35mm and 16mm high-speed film and video cameras, as well as the aforementioned high-resolution electro-optical systems. PMTC pioneered the EO systems "Cast Glance" and "Cast Eyes" that are now being utilized in MPA applications.

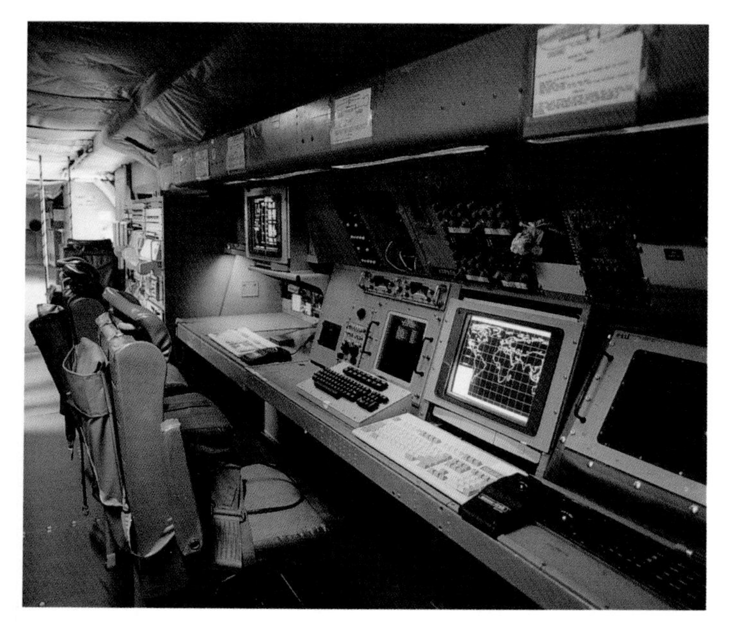

Interior of NAWC Point Mugu's new "TOMAHAWK" Orion. ***US Navy / NAWC Pt. Mugu***

Another PMTC EP-3A range aircraft (#150525) was modified as a dedicated Harpoon anti-ship missile test Orion. The aircraft is equipped with the Harpoon Weapons Test System onboard and provides testing (test firing) and evaluation of Harpoon missiles and research and development hybrids.

More recently (1995) two of the test range Orions were modified with the Tomahawk Telemetry Upgrade system. This upgrade enables two test range aircraft (#150521 and #150522) to act as command and control platforms for the Tomahawk Tactical Land Attack Missile or TLAM test programs at remote ranges worldwide without the need of costly ground control stations.

The upgrade takes advantage of the Billboard Orions' new port-side telemetry phased array antennas, housed in the horizontal extensions of the vertical stabilizers. The new upgrade includes new stand-alone mission command and control center installed aboard the aircraft. It incorporates a mission director, Navy test director and data analysis operator positions situated onboard the plane, rather than in a remote ground control center somewhere. The Tomahawk Mod

AMPS Orion. *US Navy / NAWC Pt. Mugu*

AMPS Orion. *US Navy / NAWC Pt. Mugu*

is in addition to the existing displays, recording and relay capabilities of the aircraft and in no way degrades them.

The missile's telemetry data and position information is received by the Orions' phased array and imputed into the Tomahawk thermal strip chart display and then through computer software interfaces is further displayed along with tracking information. The Tomahawk system can receive up to six data link tracks from the EATS-DataLink/GPS transceivers installed in the AIM-9 sidewinder pods attached to participating chase jet planes, or from the missile itself.

Another recent mission capability added to a PMTC Orion is the AMPS—Airborne Multi-Sensor Pod System. This is a RDT&E program to test and evaluate new technology sensors with applications towards arms control and non-proliferation of nuclear weapons materials, as well as environmental monitoring and disaster control. This multi-sensor data-collection research facilitates the possible utilization of this technology in threat verification and environmental monitoring applications.

Sponsored by the Department of Energy in association with several US national laboratories, one PMTC test range Orion (#150524) has been wired for the AMPS pods which are based on the US-3A Viking carrier onboard delivery cargo pods—each 200 inches long by 42 inches in diameter with a 90 inch wide door—that are re-configured as electronic sensor bays. The pods are mounted on the aircraft wing hard points via a specially equipped pod boot.

The initial three pods tested included a digital imaging SAR radar pod, positioned in a side-looking configuration. The second multi-sensor pod contains various optical and thermal-imaging sensors placed in a downward looking position. The third, called the "effluent species identification pod" has the capability to detect airborne radioactive particles escaping from nuclear weapons production facilities, as well as chemical pollutants and warfare agents leaking from manufacturing plants and storage facilities.

The AMPS Mod included a roll-on/roll-off twin-sensor operator stand-alone workstation installed in the back of the aircraft from which to operate the SAR pod—with controls to activate the other pods and record the incoming multi-sensor data. On their own, each pod cannot generate the degree of information that can be derived from combining data from all the sensors together. This combined sensor technology is at the heart of AMPS research to verify the accuracy of the synergism concept.

By late 1993 and early 1994, the Point Mugu based AMPS pod-equipped P-3 was flying operational evaluation flights. One series put the AMPS bird over the swollen banks of the Flint river near the town of Bainbridge, Georgia. The aircraft's mission consisted of crisscrossing the Flint river and the flooded town of Bainbridge to collect and record remote-sensing imagery for use by the Federal Emergency Management Agency (FEMA).

The river eventually crested at 37 feet, 13 feet above flood stage. With a death toll of thirty-one people killed and over forty million dollars in damage, the AMPS P-3 collected multi-spectral infrared video and SAR radar imagery of the flood area. On subsequent flights, the AMPS P-3 collected additional data that assisted FEMA in its processing of insurance claims in the area. The AMPS P-3 has also successfully imaged ground targets in and around Washington, D.C. The aircraft-collected AMPS imagery is for utilization by the US Secret Service, FBI and EPA, as well as several intelligence agencies.

PMTC, now known as NAWC-WD, Point Mugu, continues to operate its small fleet of test range Orions, although several of these aircraft have been retired and their capabilities consolidated on the remaining aircraft. Back in 1993, the

EP-3B "BATRACK" Orion. ***Lockheed***

Navy P-3 Class Desk, approved a request to re-designate all the missile test center's Orions to RP-3A—as a means to reflect the research aspects of the work conducted by the aircraft—despite varying greatly from existing RP-3A configurations both in capabilities and missions.

The new designation was short lived though, for approximately a year later, on February 1, 1995, the Naval Air Warfare Center Headquarters announced that in an effort to standardize all the various P-3 configurations at the NAWC centers, an upgrade project had been initiated to standardize the navigation and communication systems as well as safety of flight concerns of all the NAWC Orions, and upon completion of the Mod, to re-designate all the NAWC P-3s, including PMTCs, as "NP-3D" Orions.

As stated before, EP-3A configurations are varied. The EP-3A designation suggests more of a utilization in electronic test and evaluation of systems, sensors and weapons as well as airborne support to other military systems development and scientific research rather than a dedicated mission capable Orion variant aircraft as one will see further on.

EP-3B ORIONS

The next Orion variant aircraft to be developed came on the heals of EP-3A #149673, which was testing a new ELINT system, to be based on the P-3 to replace aging EC-121 Super Constellations. These new variant Orions were designated EP-3B, nicknamed "Bat Racks." There were only two EP-3B aircraft modified for the ELINT mission. The two aircraft, like #673, were modified from the remaining Black Orions used by the CIA over China. After their return to the United States and the successful proof-of-concept tests performed by #673, the two remaining aircraft (#149669 and #149678) were modified for the Navy's ELINT mission.

The ELINT systems installed on the EP-3Bs were new ship-borne systems converted for airborne utilization and relied heavily on the Black Orion's cargo door configuration to incorporate the units into the aircraft. The modification was performed by Lockheed and was completed by early 1969. The new "Bat Rack" Orions were then unofficially assigned to VQ-1 Fleet Reconnaissance Squadron in June and July 1969 under a classified project. The aircraft operated out of DaNang, South Vietnam, and flew clandestine operational ELINT flights in support of US intelligence operations throughout the rest of the Vietnam conflict and those missions assigned close to China and North Korea.

One of the reasons the "Bat Rack" Orions were designated "EP-3B" was due to the replacement of the aircraft's engines with -14 engines, as well as P-3B heavyweight landing gears. Other ELINT features of the Mod included the dorsal and ventral canoe pods, a M&M™ shaped radome mounted in the weapons bay and an extensive array of antennas bristling along the fuselage, very similar to the first EP-3A (#673) system test aircraft.

It must be recognized that there was another EP-3B not associated with the ELINT Bat Racks. This other EP-3B (#152442) was acquired by NRL in 1973 and was based on a heavyweight Bravo that was subsequently modified as a EW simulator/evaluator flying laboratory for the INEWS project. The aircraft's primary mission was to simulate hostile weapons systems to test and evaluate ship borne detection systems. The aircraft was equipped with various fleet radar, emitters and countermeasures receivers housed in wing mounted pods and onboard evaluation avionics to give instant analysis of the fleet elements being tested.

Airframe modifications included structural improvements, heavy cargo rails to accommodate roll-on/roll-off mission operator consoles and electronic racks, as well as wiring for new and existing wing hard points. Externally, additional hard points were mounted under the belly weapons bay and aft of the center wing box area to facilitate the attachment of some of the pods which were over sixteen feet long. The aircraft was also equipped with a chaff dispenser pod.

This EP-3B was additionally utilized for various other airborne projects, including electronic support measures systems and new development systems projects. More recently #442 has been deconfigured for the INEWS mission, with another NRL Orion being re-designated the primary INEWS mission platform, and re-configured as the new Airborne Surveillance Command and Control (ASC^2) development aircraft.

The ASC^2 Orion aircraft was converted, beginning in September of 1996, with a twenty-four foot E-2C Hawkeye rotodome and APS-145 airborne early warning radar. The ASC^2 project is tasked with the research and development of AEW radar upgrades, the Cooperative Engagement Capabilities (CEC) program developments and development of Ballistic Missile Theater Defense of battle groups.

Development of the CEC program provides tactical aircraft with an over-the-horizon range coverage that links surface, airborne and land based assets, thereby enhancing the battle group's anti-air warfare capability against low flying antiship missile threats.

The NRL EP-3B, recently equipped with the ASC^2 system, was re-designated as an NP-3D under the NAWC headquarters' 1995 mandated NAWC upgrade program.

Getting back to the Bat Rack EP-3B, the success of the program led to another Navy contract being let to Lockheed for ten more ELINT configured Orions. These ten ELINT aircraft went on to become the next Orion variant model to be developed, the "EP-3E Aries" Orion.

EP-3E ARIES ORIONS

The EP-3E Orions were highly sophisticated electronic reconnaissance aircraft that were utilized by fleet air reconnaissance units for tactical signal warfare support. The SIGINT Orions obtain information on targets of vital interest to the battle groups or joint tactical commander by intercepting electromagnetic signals emitted by hostile sea going vessels and or land-based installations. The EP-3Es monitor, collect, record and analyze these electronic emissions. The raw data collected can be used immediately to identify the source generating the signals or recorded for later identification and added to the EP-3E's tactical library. The EP-3E's avionics include various radar signal analyzers, directional finding indicators, intercept analysis and recording systems, ESM units and instantaneous frequency measurement equipment.

Between 1971 and 1975, ten ASW P-3As were selected from the fleet and converted into the EP-3E Orions. Unlike the modified ship-based ELINT avionics installed on the EP-3B Bat Rack Orions, the EP-3E were equipped with an all new integrated ELINT avionics. The Airborne Reconnaissance Integrated Electronic System, or "ARIES," links all the onboard ELINT system components together via a tactical processor—

NRL's "INEWS" EP-3B Orion. ***US Navy / NRL***

EP-3E ARIES Orion. ***Lockheed***

much like the advances made between the P-3A and the more sophisticated P-3C model Orion through the introduction of an integrated computer.

The new EP-3Es were externally very similar to the EP-3Bs in that they inherited the dorsal and ventral canoe pods, large "Big Look," M&M™, weapons bay mounted radome and numerous antennas. The EP-3Es were also modified with the -14 engines and heavyweight landing gears, just like the EP-3B.

It wasn't long after the new EP-3E ARIES were introduced into the VQ-1 and VQ-2 units, that the first ARIES systems upgrades began. The first started in 1978 and was known as "Deepwell." This modification included the installation of a second row of tactical workstations along the starboard side bulkhead of the aircraft, adding a "COMINT" or communications intelligence/intercept capability to the aircraft. The Deepwell upgrade also incorporated a new open-reel tape recorder and increased computer control capacity of the onboard receivers. Only seven of the existing ten EP-3E ARIES were modified with the new capability. The remaining three ARIES and two Bat Racks (which were later brought up to the EP-3E Aries configuration) were left unchanged.

The second EP-3E avionics update occurred in 1985 and upgraded some of the EP-3E receivers and replaced the open-reel tape recorders with cassette units. As before, this upgrade was reserved for the seven Deepwell aircraft only. It was also at this time that "all" the EP-3E Orions received a new AGC-9 improved teletype set for the communications suite. Some of the recent improvements were originally scheduled for the next phase of EP-3E development, the EP-3E ARIES II program, but were incorporated early.

EP-3E ARIES Orion. ***R..J. Archer via Isham collection***

EP-3E ARIES-Deepwell configured Orion. ***M. Wada via Isham collection***

EP-3E ARIES II ; note initial gray and white paint scheme with black radomes and pods. ***US Navy***

EP-3E ARIES II

With the EP-3E in continuous service since 1971, and hard time spent as ASW aircraft previously, the fatigue rate of the ARIES Deepwell Orions was steadily increasing. The Navy looked at replacement aircraft based on new production airframes, but the cost was too great. Then the Navy and Lockheed came up with a plan to modify twelve low-hour non-updated (NUD) P-3C airframes. The approved conversion-in-lieu-of-procurement program encompassed converting the twelve airframes, and standardizing the existing EP-3E Orion ELINT configurations into the aircraft with increased integration and processing capabilities between the workstations. The new ARIES II/CILOP modification also incorporated the installation of GPS and SATCOM systems into the EP-3E configuration.

The CILOP program began in 1986 with the Navy awarding Lockheed's Ontario, California, facility with a contract to engineer and manufacture the ARIES II tactical workstation components. Lockheed's Ontario AEROMOD facility in South Carolina was tasked with the stripping of operational mission equipment out of the older Bat Rack and ARIES-Deepwell aircraft and to perform the extensive re-engineering of the NUD P-3C into the EP-3E and perform the installation of mission equipment into the new ARIES II aircraft.

Due to delays in the Lockheed managed program in early 1992, the Navy CILOP contract was later re-organized and divided. The first five aircraft were completed by Lockheed AEROMOD, while the last seven NUD airframes were transferred to the NADEP at NAS Alameda for conversion. But soon after it was announced that NADEP Alameda was to take over the CILOP Mod line and the aircraft were transferred, the government's Base Realignment And Closure Commission (BRACC) slated NAS Alameda and the NADEP to close. With the BRACC-induced mandate to close Alameda and the impending transfer of P-3 depot maintenance over to the East Coast, a work force instability threatened to halt continuation of the CILOP program at NADEP Alameda.

Interior tactical compartment of EP-3E ARIES II; portside forward (crew stations 8-14) ***Marco Borst***

EP-3E ARIES interior; starboard side aft (crew stations 15-19) ***Marco Borst***

EP-3E ARIES interior; secure communications (crew station 4)
Marco Borst

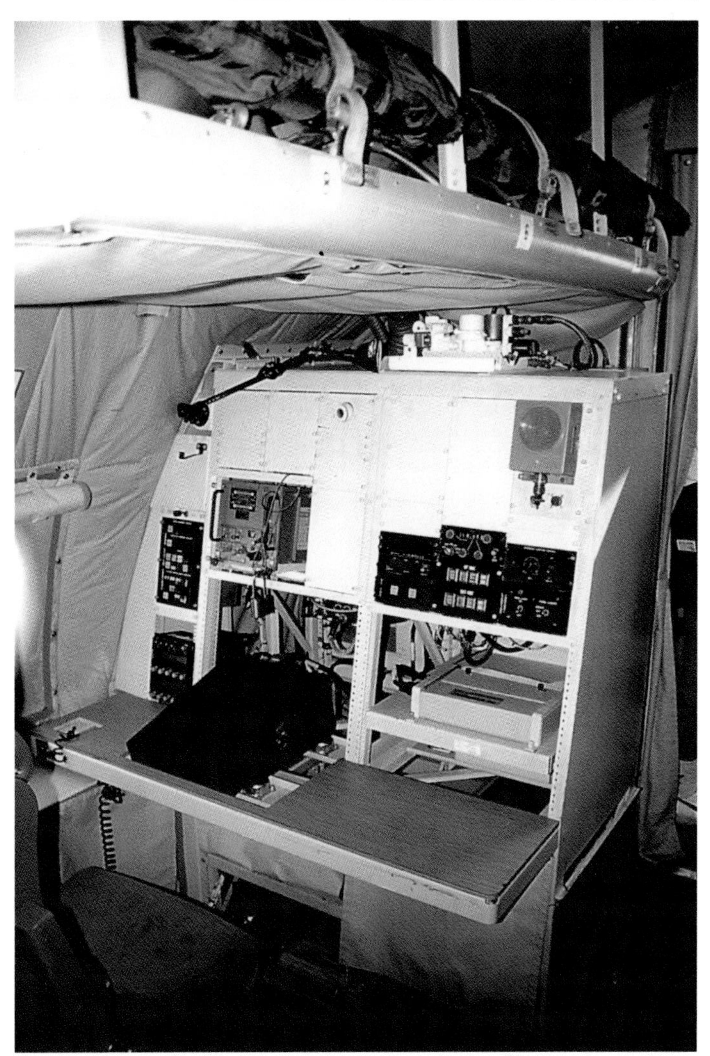

EP-3E ARIES interior; scientific /tech operator (crew station 20)
Marco Borst

After much debate, the Navy decided to split the Mod line again and transferred conversion of the remaining three EP-3E ARIES II baseline aircraft to the NADEP at NAS Jacksonville, Florida. NADEP Alameda would continue to work on the four CILOP aircraft that would be delivered through December 1996. The last three Charlie baseline aircraft were inducted into the NADEP Jax Mod shop in 1995. Prior to induction, each airframe received a standard depot maintenance rework. Delivery of the remaining EP-3E ARIES II out of NADEP Jax ran through 1997.

As the Cold War faded and the US Congress began looking for areas to cut the Defense Budget, the future use of the EP-3E came into question. A matter of mission redundancy between the capabilities of the EP-3E and the USAF's RC-135 tactical intelligence aircraft became an issue in the minds of Congress. In Fiscal year 1993, CILOP funds were frozen and the program was halted, contributing to delays at Lockheed.

A study was initiated to determine which airframe was to be considered for future upgrading and carry on as the primary signal and electronic intelligence gathering aircraft and which one would be phased out. This eventuality was contrary to the belief of many in the Pentagon that each aircraft had its own unique capabilities and that they complimented each other. The study, later presented to Congress, highlighted the EP-3Es' unique contributions to the recent Gulf War with Iraq. It was there that the EP-3E proved their versatile mission capabilities.

As tensions escalated, the EP-3E assumed their roles in conducting electronic reconnaissance, threat indications and warning missions for coalition forces and Arabian Gulf based battle groups, with detachments from VQ-1 and VQ-2 operating out of Bahrain. The EP-3Es were also tasked with strike reconnaissance and Bomb Damage Assessment for "HARM" missile strikes into Iraq. Some of VQ-2 EP-3E, stationed in Souda Bay, Crete, provided threat warning surveillance to

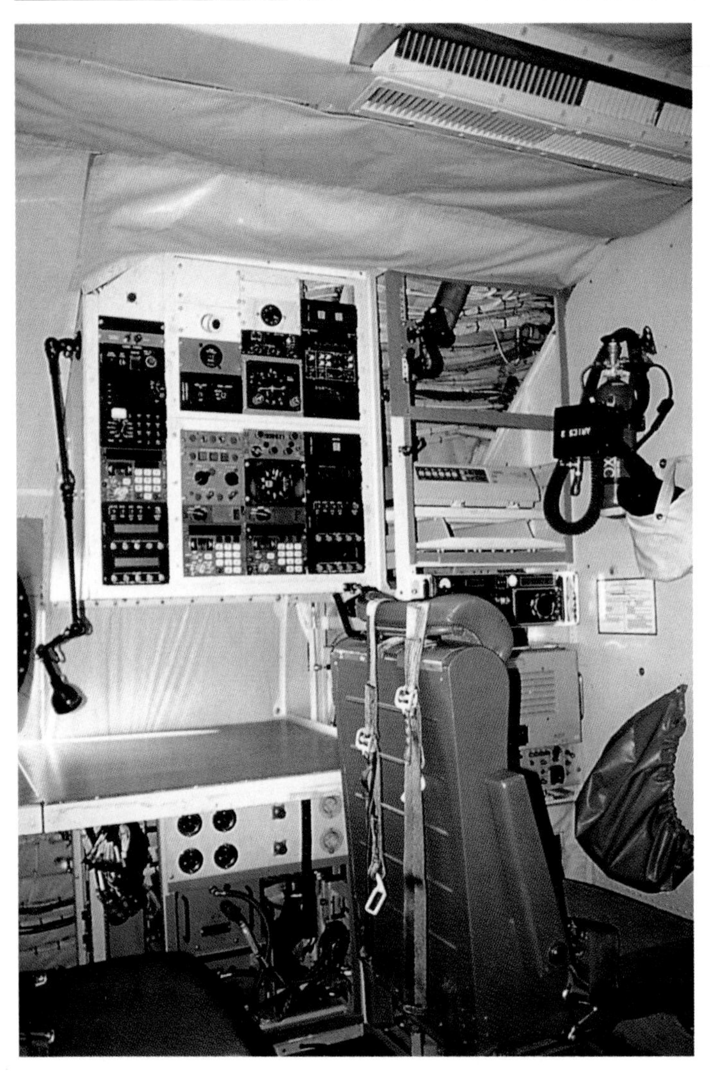

EP-3E ARIES interior; navigator position (crew station 6) *Marco Borst*

coalition forces and strike reconnaissance for USAF B-52 sorties out of Turkey into Northern Iraq. These EP-3Es later flew surveillance missions in support of "Provide Comfort" humanitarian relief efforts for Kurdish refugees along the Iraqi border with Turkey.

Subsequently, the US Congress agreed that the EP-3E and RC-135 missions were complimentary and unfroze CILOP money to resume EP-3E ARIES II modifications.

As the EP-3E ARIES II modification process continued, a follow-on upgrade was being developed. Designated "SSIP" for Sensor System Improvement Program, the upgrade incorporates increases in frequency and directional finding coverage across the board via state-of-the-art systems. SSIP is to be implemented in two phases: Phase I targets the most important systems needing immediate upgrading to fulfill immediate requirements, such as those effecting joint interoperability and communication conductivity of the aircraft. This includes installation of new communications and link systems to standardize the aircraft with the rest of the fleet and the joint tactical arena. Phase II will take advantage of new emerging technologies to enhance its mission capabilities.

SSIP equates to heightened situational awareness for the mission commanders through auto-correlation of organic tactical information (data the aircraft is generating) with non-organic data (that which is received from new tactical Link and communications networks). This is accomplished through the "STORY" series of system enhancements to improve aircraft interoperability with other airborne reconnaissance/surveillance platforms, satellites and ground-based command and control centers.

The STORY series includes: "STORY TELLER" for the automatic organic /non-organic all-source data fusion and enhanced communications connectivity via new communication Link networks; "STORY BOOK" adds the means to

EP-3E ARIES II ; note gray and white paint scheme with gray radomes and pods. *US Navy*

EP-3E ARIES II today ; all gray with gray radomes and pods. ***US Navy via J. Turnbull***

exploit old and new digital data Links and data fuse them into the aircraft's organic mission systems; "STORY CLASSIC" upgrades common crypto logic workstations, incorporating increased signal detection capabilities across the board with greater systems reliability. SSIP will also add a ULQ-16 pulse analyzer upgrade for greater real-time, on-board signal analysis.

Introduction of SSIP into the EP-3E Aries II begins with the last two EP-3E CILOP aircraft off the NADEP Jax Mod line. The remaining ten aircraft will be upgraded by the end of the century.

One aspect of the ELINT Orion variant is its paint scheme. The original EP-3B "Bat Rack" Orions were delivered in midnight blue/black and white, with black radomes and pods. Later they were painted gray and white (with black radomes and pods) when upgraded to EP-3E ARIES capability. As far as the EP-3E ARIES II, there were some questions as to whether the original CILOP specifications stipulated a particular paint scheme other than the traditional EP-3E gray and white. It seems that the CILOP specifications meant for the ARIES II aircraft to be painted in an all-gray tactical paint scheme. New composite material radomes and canoe pods were delivered all-gray.

But Lockheed Aeromod produced EP-3E ARIES which were delivered gray and white with the old black radomes and pods off the ARIES-Deepwell donor EP-3E. This apparently deviated from the design specifications. The fifth Lockheed ARIES II actually was delivered gray and white with gray radomes and canoe pods. The first four ARIES II were later refitted with the new gray fixtures in the field.

The NADEP Alameda-produced EP-3E Aries II were rolled out all tactical gray with gray pods and radomes, as were the NADEP Jacksonville EP-3E Aries II. The suggestion is that the all-gray paint scheme is as the specifications intended and that the new radome and pods were gray for that reason. In any case, all EP-3E Aries IIs will subsequently be repainted all-gray during future SDLM maintenance.

WP-3A ORIONS

Another dedicated mission capable Orion was the next P-3 Variant to be developed by the US Navy. The new Orion was designed for the art of Hurricane Hunting as part of a humanitarian service tasking performed by the Navy since the middle of World War II.

Airborne weather (hurricane) reconnaissance was born during WWII and established by both the US Army Air Corps (now the US Air Forces) and the US Navy in response to hurricanes and typhoons that complicated sea going task forces and land based combat operations in the Atlantic and Pacific theaters of operation during the war. In one case,

WP-3A weather reconnaissance Orion. ***Isham collection***

WP-3A Orion. ***K. Buchanan via Isham collection***

Admiral Halsy's Pacific task force was severely crippled by typhoons with the lost of several ships during critical operations against forces of the imperial Japanese Navy.

Both the services took the lead in 1944, establishing weather reconnaissance units and modifying existing patrol and bomber aircraft into "Hurricane Hunters" and "Typhoon Chasers." Between 1945 and 1958 most reconnaissance included periphery flights to locate storms and fly around the storm collecting meteorological data to determine the speed and heading of the tropical storms and hurricanes. It wasn't until 1948 that the introduction of "airborne radar" into hurricane hunting aircraft occurred and provided the means to peer into the violent torrents to unlock their mysteries. Several years later the first (storm) eye penetrations occurred, which provided the means to collect wind and pressure measurements to asses the strength of a storm.

During this period, a number of military reconnaissance units were formed and evolved. One Navy unit went from being designated patrol bomber squadron 114 (VPB-114), tasked as the first Navy weather reconnaissance unit, to Weather Squadron 3 (VPW-3), to Meteorological Squadron 3 (VPM-3), and finally to Heavy Land-Based Patrol Bomber Squadron 3 (VP-HL3) by 1946. But this was a part time seasonal reconnaissance unit, as was VP-23, the Anti-Submarine Patrol Squadron.

Operating out of Naval Air Station Miami, the Squadron performed hurricane reconnaissance in the Caribbean until season's end and then deployed north to NAS Argentia, Newfoundland in Canada for ASW patrol. This unit was later transferred to the maritime patrol base in Brunswick, Maine, leaving behind a hurricane reconnaissance detachment which later became a dedicated hurricane reconnaissance squadron, designated VJ-2. It was VJ-2 that, after a transfer to the Naval Air Station in Jacksonville, Florida, in 1952, became the "famous" VW-4 hurricane hunters of the US Navy. A unit that would eventually acquire the WP-3A Orion.

VW-4 (and its predecessors) have operated a variety of aircraft over the years encompassing PBM-1 Mariner seaplanes, PB4Y-2 Privateer Patrol Bombers, P2V-3W/P2V-5JF Neptunes, WV-1 Constellations and WV-3 Super Constellations (later re-designated WC-121N) prior to the subsequent acquisition of the WP-3A in 1970.

In late September 1969, the Navy was afforded an opportunity, with the appearance of Hurricane Inga that tracked up off the southern area of Bermuda, to assess the capabilities of proposed follow-on weather reconnaissance aircraft for VW-4's WC-121N Warning Star aircraft. The Naval Air Test Center's "Service Test Division established a qualification flight test program to operationally test candidate aircraft to determine which one would be suitable to carry on the Navy's weather reconnaissance mission. The proposed airframes included a newer, upgraded version of the Lockheed WC-121N, a Lockheed C-130 Hercules and a P-3A Orion.

NATC flew the aircraft in an unprecedented three plane trailing formation with a fifteen mile separation. Over twenty low level penetrations into Hurricane Inga were made. The service test division evaluated the aircraft in various parameters consisting of propulsion systems and flying qualities and handling performance, as well as structural integrity and payload capabilities to carry reconnaissance equipment and reserve power. Human factors such as temperature, humidity and air conditioning in the working environment were also evaluated as well as fatigue, safety and quietness parameters.

Although all the aircraft performed very well, the US Navy selected to base its follow-on weather reconnaissance platform on the P-3 Orion. It's interesting to note that the US Air Force has been utilizing C-130 Hercules as Hurricane Hunters since 1960.

The new weather reconnaissance Orions, designated WP-3As, were developed from existing ASW P-3s within the Fleet. Under a Navy contract, Lockheed's Ontario, California, rework facility modified the Orion aircraft into the Hurricane Hunters that they would become. The Mod program began with stripping out all the old ASW gear and sensors from the aircraft. The aircraft was then modified with a ventral weapons bay mounted, M&M type radome housing an APS-20 radar. The WP-3A would also retain their APS-80 radar in the nose and tail radomes and received structurally strengthened wings. Mission equipment included the DALS, or Data Acquisition Logging System, which simultaneously collects, records and transmits data over radio frequencies back to the National Hurricane Center. Aerographical instruments installed aboard the aircraft encompassed Vortex Thermometers, for measuring DEW points, dual radar altimeters and a Bathythermal sonobuoy receivers and launch system, not to mention a Wind-Dropsonde system, which radios back vital pressure and meteorological data from the heart of the hurricane once dropped via the aircraft's free-fall tube. The WP-3A had one of the most extensive navigation suites of the day with dual inertials, Doppler and Omega navigational systems.

The WP-3A interior layout remained practically the same with the tactical compartment re-equipped with new mission operators. The side-by-side bench was re-configured with, looking aft to forward: the BT Recording/Dropsonde Receiver Operator station; the Navigator's Station and next to it, two Radar Operator Stations—one for the APS-20 and the other for the APS-80 radar system.

Additionally, just behind the cockpit on the port side, in the old radio operator's station, was the Mission Weather Meteorologist position. The WP-3A also retained the P-3A's aft observer stations for visual observations, as well as additional seating throughout the aircraft.

The WP-3As were capable of collecting and recording airborne meteorological "weather" information, including wind speed and Barometric pressures used to determine the speed, direction and strength of tropical storms and hurricanes. The aircraft were also capable of gathering basic oceanographic data, as well as a platform from which to conduct limited weather or meteorological "Research," such as the infamous "Project STORMFURY" experiments.

Project STORMFURY was a DOD/Department of Commerce storm seeding research project to see if dropping Crystal/Silver Iodine into the clouds of tropical storms and hurricanes could take the punch out of these destructive storms. The WP-3As were just one of a number of aircraft used for storm seeding missions. The WP-3As were equipped with a 40 mm rocket launcher like device mounted on an out-board hard point on the port wing.

Project STORMFURY was eventually canceled due to many reasons, the primary reason being that seeding often caused the storm direction or size of the storm to change, causing more problems than helping. Geopolitically, if these storms could be controlled, others would see the US as vying to control the weather as a possible weapon. It's interesting to see that shortly after the STORMFURY project began, the then Soviet Union began looking at hurricanes and typhoon reconnaissance/research that led them to develop a couple of specialized "Hurricane Hunting" aircraft that had a capability of meteorological weather modification.

Besides seasonal hurricane reconnaissance missions, VW-4 and its WP-3As were tasked with various missions around the globe. These additional mission taskings encompassed Northern Atlantic Polar Ice Reconnaissance, East coast Winter Storm Warning (a sort of hurricane hunting of winter storms that often paralyze the US East coast in winter), the Indian Ocean Oceanographic Reconnaissance tasking and the Labrador Current Sea-Surface Study, as well as bi-weekly meteorological and oceanographic deployments over the North Atlantic and Mediterranean in support of the fleet operations.

RP-3 Orions; Projects MAGNET, SEASCAN and BIRDSEYE. ***US Navy***

RP-3D Project MAGNET. ***Lockheed***

One project, the Hydrometer Erosion And Retardation Test, or "HEART," was assigned to VW-4 and required the squadron to specially modify their logistical bounce P-3A to conduct the Project. Although not designated a WP-3A, the specially configured VW-4 Orion measured the parameters of ice crystals' destructive effects on the nose cones of re-entry missiles or satellite pods, which cause disruptions in planned trajectories.

In late 1974, in the wake of the then Carter Administration military budget cuts, the US Navy announced that VW-4 would be disestablished and the WP-3A retired from service—turning over thirty years of history in humanitarian relief operations to the USAF and their hurricane hunting C-130. On April 30, 1975, the Hurricane Hunters of VW-4 were decommissioned and the weather reconnaissance Orions retired. Actually, the aircraft themselves weren't really retired at all, but were re-engineered into other P-3 variants. One aircraft (#149674) became an EP-3A with NRL and the others (#149675, #149676 and #150496) were converted into another Orion variant—the VP-3A.

RP-3 ORIONS

Between the initiation of the WP-3A Orion and another Orion variant called the VP-3A, which was developed from ex-weather reconnaissance P-3 airframes, the US Navy established the RP-3 Orions. The new Orion derivatives were developed for world-wide oceanographic research and data collection in support of Navy fleet operations. Despite having an "R" designation, usually reserved for intelligence gathering reconnaissance aircraft, the Navy RP-3s were non-hostile research platforms dedicated to airborne oceanographic and geomagnetic surveys—hence their unique international orange and white paint schemes to easily identify them as overt non-combatant research aircraft with peaceful intentions.

RP-3D MAGNET. ***Georges Van Belleghem***

RP-3A Project Birdseye. ***R..Leader via Isham collection***

The RP-3s were produced as replacement aircraft for a specialized Navy unit designated Oceanographic Development Squadron Eight, or VXN-8. The squadron was established as VX-8 in 1967 and was tasked with various oceanographic missions and special projects such as the ASW Environmental Prediction System and Project "JENNY"—a special tasking providing airborne broadcasting of radio and television programs to the Vietnamese people during the Southeast Asian conflict. The Squadron, re-designated VXN-8 in 1969, operated NC-54 Sky-Master and NC-121 Super Constellations that were replaced by the RP-3 Orions starting in 1973.

Actually, RP-3 development began in 1971 when the US Navy took delivery of the one and only production version of the RP-3D Orion. This RP-3D aircraft (#158227) was the 51st P-3C off the Lockheed production line. The Charlie Orion airframe was specially modified on the production line with a non-ferrous, magnetically neutral tail section designed to house a highly sensitive magnetometer for VXN-8's "Project Magnetic." Project Magnetic was a geomagnetic airborne survey program collecting data and measuring the strength of the earth's magnetic fields. The new specialized Orion was the fifth aircraft to perform the Project Magnetic mission, proceeded by P2V-2 Neptunes, a NC-54 Sky-Master and a NC-121 Super Constellation.

The production RP-3D configuration varies from the P-3C and other RP-3 configurations through the incorporation of a Geomagnetic Airborne Survey System, or "GASS" magnetometer, in the non-magnetic tail emplinage. The main cabin door was also magnetically clean to provide distortion-free operation, enabling the sensitive vector airborne magnetometer to collect worldwide geomagnetic data required to form

RP-3D "dual-mission" Orion (Project Birdseye) ***Terry Taylor***

RP-3D "dual-mission" Orion (Project Seascan) ***D. Remington via Isham collection***

accurate isomagnetic charts. These charts were used to calibrate MAD systems aboard fleet ASW Orions and help ship and aircraft navigation. The RP-3D production airframe interior configuration encompassed a P-3A/B tactical workstation (bench) with two GASS operator positions and two navigators. The NAV stations were equipped with an extensive array of navigation systems including Doppler, multiple GPS, LORAN C and ADL-21 auto-track Loran NAVSET, as well as ring-laser gyro Inertials.

The aircraft's interior is further varied from the Charlie airframe in the location of the galley, which is forward of the operator's tactical workstation. There was additional seating and a second dinette set up that turns into crew rest bunks in conjunction with overhead bunks in the approximate area of the P-3C's TACCO station. The aircrafts' communications suite/operator station is located across the aisle in the Charlie's NAV/COM position. There were four additional bunks, a head and seating/auxiliary sleeping area in the back of the airframe forward of the magnetometer.

The "MAGNET," Orion's airborne geophysical survey system avionics, include the AGQ-81 Magnetometer, two Honeywell 316R digital computers and AGC-6TTY magnetic tape unit, as well as a APS-115 search radar in the nose.

Old RP-3D MAGNET; now a NP-3D Orion with NRL. ***US Navy / NRL***

The fourth RP-3D version Orion; now a NP-3D with NRL. ***US Navy / NRL***

The production RP-3D was also equipped with a sixth fuel tank permanently mounted into the aircraft's weapons bay. This gave the aircraft an extended long range capability. It enabled the aircraft to set a 5,461 nautical mile non-refueled flight, a closed-circuit distance record in 1972 for an aircraft of its type.

The Project MAGNET mission was established in 1951 to collect worldwide geomagnetic data along predetermined paths, managing and recording the data on digital computer tape for later analysis. The earth's surface magnetic field is in constant change and must be continuously updated. The GASS gear was so sensitive that solar activity often disrupted the geophysical magnetic readings and for that reason the majority of the MAGNETIC flights were flown at night.

After its outfitting with mission equipment by the Navy, the MAGNET RP-3D was delivered to VXN-8 in 1973 at its NAS Pax River (Maryland) base and was soon conducting mission systems tests. It was during this period, while flying off the coast of North Carolina, that the MAGNET Orion discovered the wreckage of the long lost Civil War Ironclad Warship the "MONITOR." A project known as operation "CHEESEBOX" was established, in conjunction with the Naval Oceanographic Office and the US Naval Academy, to locate the Civil War ship. After the location of the Ironclad was

The old dual-mission RP-3D Birdseye "Arctic Fox"; now a NP-3D Orion with NRL. ***US Navy / NRL***

NRL NP-3D in INEWS configuration. ***US Navy / NRL***

spotted on geodetic tapes collected during an earlier high-altitude systems test flight, the new analysis revealed that the Ironclad was within a few hundred yards of the earlier track contacts. The Monitor has since been salvaged by a civilian company with artifacts put on display. The Project MAGNET RP-3A also magnetically monitored the birth of a submerged volcano that subsequently erupted up out of the sea near Iceland. This volcano went on to form the new Icelandic island of "SURTSEY."

RP-3A

VXN-8 was the only Navy squadron devoted solely to oceanographic and geomagnetic surveys. The squadron was tasked by NAVOCEANO, headquartered in Bay St. Louis, Mississippi. Civilian scientists often accompanied the Navy aircrews around the world to assist in data collection and interpretation. It was NAVOCEANO that tasked the squadron with performing other oceanographic project missions to collect vital oceanographic data in support of fleet deployment operations, military plans and ASW missions.

One of these projects was called "BIRDSEYE." This project was initiated in 1962 in response to the Navy's (fleet) requirements for continuous iceberg and pack ice reconnaissance and survey to develop capabilities of predicting Arctic basin and marginal zone oceanographic environmental conditions in support of Arctic Ice ASW operations. Under this project, visual and instrument data collection and observations were made and included studying ice formations to determine formation types, distribution and boundaries in the harsh environment of the Arctic Circle. This information was beneficial to Arctic ASW, Undersea (ice) Warfare and logistical operations of fleet elements.

Another of VXN-8's projects was "SEASCAN." This project studied the ocean environment to gather vital data on ocean currents, wave heights and surf conditions, sea floor topography, algae masses and deep sea thermal layers and acoustic propagation (ambient noise generation, sounds indigenous to a particular body of water)—all of which affect ASW detection. This ocean survey data was essential to fleet ASW operations world wide.

NRL NP-3D in "LADS" configuration. ***US Navy / NRL***

Since 1973, VXN-8 had operated specialized Orions for the Birdseye and Seascan projects designated RP-3A. These aircraft were modified from ASW P-3As that were stripped of all ASW equipment and re-equipped with specialized mission instruments. Although the RP-3As were modified individually for their respective missions, the original RP-3A shared a similar data management system known as the Airborne Data Acquisition Precision System. Other common RP-3A avionics encompassed an AAR-52 Sonobuoy Receiver Set, PRT-5 Laser Profiler and a precision radiation Total Temperature Probe, as well as a Dew Point Hydrometer. Both aircraft were equipped with a variety of sophisticated navigational systems for that time, such as a GPS and very low frequency Omega navigation system.

US Navy VP-3A Orion. ***US Navy***

VP-3A Orion. ***F. MacSurley via Isham collection***

It must be noted that this configuration was originally developed for VXN-8 and was not similar to any other P-3 aircraft that at one time or another were re-designated RP-3A temporarily.

The first RP-3A (#149670) was assigned to Project Birdseye. This aircraft was later retired by VXN-8 and was re-engineered again as a EP-3A test aircraft for NRL. The second RP-3A (#149667) was assigned to Project Seascan. This aircraft was also later retired by VXN-8 and was replaced by another P-3A (#150500) that was modified and converted into an Birdseye RP-3A.

In 1990, both the existing project Birdseye and Seascan RP-3As were replaced by two RP-3Ds, which had been converted from a heavy weight P-3 Bravo and the original YP-3C prototype Orion.

Although the P-3B (#154587) was modified into an RP-3D for the Birdseye mission and the prototype P-3C (#153443) was converted for the Seascan Project, both aircraft were capable of performing the other's mission. These conversion RP-3Ds were dual-mission capable. They were re-configured by the Naval Aviation Depot in Jacksonville, Florida.

There were two other RP-3D designated aircraft utilized by VXN-8 prior to delivery of the new duel-mission aircraft from the NADEP Jax Mod shop. They were two P-3 Bravos (#152738 and #154600) designated RP-3D and minimally configured to perform both the Birdseye and Seascan missions. They were called "mini-mod" aircraft that were utilized as interim mission platforms while the dual-mission RP-3Ds were undergoing modification. These mini-mod aircraft were provisioned with rail-mounted racks for roll-on/roll-off mission electronics, including GPS.

The VXN-8 RP-3 Orions were known worldwide, partly because of the cartoon characters painted on the forward fuselage. These cartoons included the "Roadrunner" on the Project Magnet aircraft (nicknamed Pisano Tres), "El Coyote" on the Seascan RP-3 and "Arctic Fox" on the Birdseye Orion. The newest dual-mission Birdseye RP-3D was actually the fourth P-3 Orion to bear the Arctic Fox cartoon over the years since 1973. There were the two RP-3A mission aircraft, one UP-3A auxiliary back-up aircraft (#151384) and the newer dual-mission RP-3D.

VXN-8 also operated a number of UP-3A Orions for logistical and training aircraft and they too bore decorative cartoons. One was the "Tasmanian Devil" (#150527) and the other (#150528) was nicknamed "Loon" with a loon cartoon. All these cartoons were Warner Brothers developed and sanctioned characters to be painted on the Orion aircraft.

Navy budget cuts, in the wake of the Cold War, have had their effects on the P-3 community and VXN-8 was no exception. Since 1987, VXN-8 had been officially considered the twenty-fifth active maritime patrol squadron rather than just a special mission unit. Although a prestigious honor for the squadron at the time, it seems to have spelled its demise. The unit was commanded to stand down by October 1993, due to lack of funding in the fiscal year 1994 budget.

Since the announcement, NAVOCEANO began formulating options as to what to do with the RP-3Ds and who would take over their missions. Considerations ranged from base commands such as NAS Pax River or NAS Brunswick Maine to absorbing the mission taskings into a regular fleet patrol unit or reserve patrol squadron (VP-94 was mentioned due in part to its close proximity to NAVOCEANO), or to let out a government contract to a civilian operator to take over the VXN-8 missions and aircraft. In the end, the Naval Research Laboratory's "Flight Support Detachment" was selected and took delivery of RP-3D's Arctic Fox (#154587) and Roadrunner (#158227) aircraft. The El Coyote RP-3D (#153443) was transferred to NAWC Pax River to be utilized as a range test aircraft.

With NRL, the Birdseye RP-3D has been wired with fiber-optics to provision the aircraft for the INEWS mission as a back-up aircraft for another NRL RP-3D that has recently been selected as the new primary INEWS aircraft. The Birdseye aircraft will also absorb and perform the Seascan mission besides its own ice reconnaissance tasking.

It's interesting to note that there is another RP-3D with NRL. Like the dual-mission research P-3 produced for VXN-8, this aircraft (#154589) is also based on a heavy weight Bravo airframe converted by NADEP Jax. The aircraft had been acquired by NRL in 1990-91 for a proposed project that would have had it re-engineered with a 24 foot rotodome and radar system from an E-2C Hawkeye. NRL stripped out the aircraft and had plans to install the rotodome, but funding was cut and the conversion was scrapped. Later NRL decided to utilize the aircraft as a back-up mission platform for the EP-3B tasked as the primary INEWS plane. It became the first P-3 Orion to be equipped with fiber-optics that replaced a ten to twelve inch thick bundle of wiring installed in the leading edge of the wing to run the INEW pods. The aircraft was also fitted out as a general NRL project test aircraft which included interior power and cooling outlets throughout the aircraft. This aircraft was originally scheduled to be designated as another EP-3B, but once complete it subsequently became the fourth version of the RP-3D.

This aircraft recently tested a new airborne laser system in Australia. The Laser Airborne Depth Sounder system, or "LADS," is a Australian laser system used to conduct bathymetric surveys of underwater topography. The LADS unit was housed in the Orion's bomb bay and the operator station was fitted into the back of the aircraft.

The NRL LADS project was to collect vital operating parameter data required to assist US Navy ASW laser systems planned as new P-3 operational sensors in the future. The NRL P-3 fleet is currently capable of carrying a portable P-3 bomb bay pallet system framework. This is a bomb bay mounted rack system that allows for a means easier of testing R&D electronic sensor systems. A system under development can be fitted into the pallet framework which is then mounted into the weapons bay of most NRL Orions. The versatile Mod also includes fairings installed in place of the bomb bay doors.

As for the original RP-3D, Project MAGNET, the future is in doubt. Current options include stripping the aircraft of its GASS mission equipment and converting it into a multi-mission test platform. The mid and high altitude geomagnetic data is no longer conducted by the aircraft, but is collected by the "MAGSAT" satellite. Old MAGNET could also be retired all together. NRL is still trying to decide which mission best fits the aircraft.

Since the RP-3Ds' assignment to NRL, the aircraft have been re-designated NP-3D on February 1, 1995 under the NAWC upgrade program. This upgrade addressed the navigational, communications and safety of flight concerns of NAWC P-3 Orions. Upgrades to the aircraft have included satellite communications and new radios, as well as additional navigational gear (if needed) and color weather radar.

Due in part to NRL acquiring VXN-8's Orions, the unit is planning to repaint its remaining Orion fleet in the non-hostile international orange and white paint scheme reminiscent of the old VXN-8 RP-3 Orions. Although the RP-3 Orions may no longer exist, their essence lingers on!

VP-3A

As time progressed another P-3 variant was developed by the Navy. This group of P-3 derivatives were extensively re-engineered as VIP transport platforms for military dignitaries. In 1975, with the availability of ex-service WP-3A Orion airframes from VW-4 Hurricane Hunters, the Navy established the requirement for new VIP transport aircraft. Three of the surplus WP-3As (#149675, #149676 and #150496) and two fleet ASW P-3As (#150511 and #150515) were selected and converted into new VP-3A Orions.

The conversion work was conducted by NADEP Jax and entailed completely stripping out all of the sensors, operator stations and any equipment. The interiors were then refurbished with executive suite facilities, staff seating (area), conference wardroom and a private stateroom with lavatory. A number of storage bins and cabinets were incorporated along the interior of the fuselage where possible. Amenities consisted of sidewall wood paneling, plush carpeting and first class airline seats. The Mod included an upgraded, fully functional galley in the aft tail section of the aircraft. This new galley refit replaces the austere food preparation and eating area of the ASW sub hunting Orions. These galleys included full size appliances such as a large size refrigerator, stove and microwave oven. The VP-3As are basically laid out with the wardroom forward of the galley—across from the main cabin door—with the Admiral's private stateroom forward of the wardroom. Additional staff seating is located still further forward and where ever possible.

Not all the VP-3As are configured the same. Some of the aircraft have slight differences in regards to the size of the executive cabin, staff seating area and location of the galley or quality of the interior appointments.

The VP-3As also incorporate upgraded avionics including state-of-the-art navigation systems, such as GPS and sophisticated communications systems comprised of new VHF radios, satellite communications and data link capabilities for military command and control missions. The VP-3A are also equipped with color weather radar.

Three of the VP-3A Orions are operated by the Administration Support Aircraft (ASA) division of VP-30 based at NAS Jacksonville. Another VP-3A is located at NAS Barbers Point with the station's Executive Transport Department. A fifth VIP Orion aircraft flies with a VR Detachment based at NAS Sigonella.

CP-3A

When considering the different derivative Orions developed over the years, it's easy to get caught up in all the P-3 variants proposed by Lockheed that never got past the drawing board. One included a P-3 long range weapons platform for "PHOENIX" missiles and another encompassed re-inventing

TP-3A Orion. ***US Navy***

TP-3A Orion. ***G.Lebaron via Isham collection***

the P-3 back into an airliner/passenger transport aircraft for Taiwan.

One proposed variant that "is" worth discussing is the "CP-3A." In 1984 the US Navy let a contract to Lockheed to produce a development program to convert twelve maritime patrol ASW P-3As into CP-3A Advance Base Support Aircraft. These CP-3As were to have been used for military cargo and passenger transport in support of Navy logistics. The modification program was to have included stripping out of all the ASW sensors and equipment before structurally strengthening the floors and installing cargo and passenger seating tracks.

The modification was also to have added a clam shell cargo door much like transport configured L-188 Electras. Additionally, the CP-3A airframe was to have been equipped with several extra emergency exits throughout the aircraft, as well as receive upgraded navigational systems and a new communications gear installed into the cockpit.

A prototype CP-3A was to have been produced and delivered by October of the next year (1985), but unfortunately the Navy canceled the contract and the program was discontinued.

This period marked somewhat of a turning point in variant aircraft development, where the Navy began utilizing its own aircraft rework and repair facilities to begin creating derivative Orions. The termination of the CP-3A program, presumably due to cost, would eventually lead to the Navy developing its own Logistical transport version of the P-3.

TP-3A

As the Navy was working out its budgetary problems with the cancellation of the CP-3A Orion program, it was considering a requirement from the VP Fleet Replacement (training) squadron for a new dedicated P-3 pilot training aircraft. What was needed was a P-3 training aircraft with a P-3C configured cockpit and devoid of all the sophisticated and expensive ASW equipment.

The Navy proposed the "TP-3A" variant to be converted from twelve available maritime patrol ASW airframes with the work performed by the NADEP in Jacksonville. The Mod program encompassed the removal of all ASW avionics and equipment including MAD, ESM, radar and sonobuoy launch components and plumbing. The tactical workstations were retained, but fitted with blank-out panels over all the mission sensor displays and controls that were removed.

After stripping out unwanted mission gear, the flight station was re-configured and upgraded to the P-3C Update II.5 configuration. New navigational and communications systems were installed, as well as a color weather radar unit and display. Besides the interior changes, a heavyweight-soft landing gear change was also performed.

The TP-3A Orion facilitated Improved pilot and flight engineer transition training and saved wear and tear on fully mission capable P-3Cs. Eleven TP-3As (out of the twelve produced) were split between Fleet Replacement Squadrons VP-30 and VP-31. The FRS units were tasked to provide qualified pilots, flight officers, aircrew and maintenance personnel to P-3 patrol squadrons.

In fiscal year 1993, VP-31 based at NAS Moffett Field was disestablished and consolidated VP training operations at VP-30 onboard NAS Jacksonville. By 1995, half of the TP-3As had been retired to the AMARC desert aircraft storage facility in Tucson Arizona. The remaining training Orion derivatives have recently been selected for replacement by a new proposed training aircraft—the TP-3C.

TP-3C

The TP-3C Orion program is a proposed US Navy modification plan to augment current P-3C force levels by providing CONUS pilot training/replacement aircraft to individual patrol

Proposed TP-3C Orions. ***US Navy via VPI***

squadrons where needed, especially when their mission aircraft planes are involved in AIP or SRP programs. The original modification plan encompassed stripping out approximately 36 P-3 heavy weight Bravos, much like the TP-3A, and upgrade the cockpits to the P-3C Update III configuration and the addition of floor tracks for logistical seating. The TP-3Cs are expected to be utilized as pilot proficiency training bounce-bird aircraft, replacing older TP-3A training aircraft with VP-30 while others are to be pooled at the wing level for use as bounce and logistical support aircraft by VP squadrons on an as-need basis. The program is to begin in 1996 pending funding.

In the meantime, a number of non-updated (NUD) P-3C have been identified to be re-designated TP-3C Orion. Under this interim measure, the NUD P-3C will be stripped of equipment and mission systems and be utilized as pilot training aircraft at the remaining VP-30 FRS unit. This is a temporary measure until conversion funding for the P-3 Bravo to TP-3C program can be procured. The select NUD P-3C to be converted will eventually be reclaimed by the P-3 program office and be re-configured into Update III.

UP-3A

It was in 1985 that the Navy initiated a low cost alternative to the logistical transport Orion proposed by the previously contracted CP-3A. This new Orion variant was designed by NADEP Jacksonville, who at the time was engaged in modifying P-3As into TP-3A Pilot trainers. What the NADEP proposed was a utility version of the P-3A, that once stripped of all avionics and ASW equipment, including the tactical workstation bench and electronic racks, could be equipped with floor tracks for logistical cargo hauling and passenger seating/transport.

NADEP was then authorized to produce a prototype aircraft. It chose to take the twelfth and final TP-3A (#151367) undergoing modification and run it back through the NADEP Mod shop. At that point, they removed the tactical workstation bench and electronic racks before installing floor cargo/seating tracks. Completely filled with passenger seats, the aircraft resembles something like the airliner it was originally derived from.

NADEP Jax unofficially designated the prototype utility Orion a "UP-3S" for standard utility P-3 configuration, but the aircraft was subsequently officially designated UP-3A. A number of ex-ASW P-3As would later be converted into UP-3As.

Although the UP-3A Orion was to have principally been a logistical cargo/passenger transport alternative to the CP-3A, the utilitarian nature of this aircraft, as suggested by its des-

UP-3A transport Orion. M. K. ***Miller***

UP-3A base transport Orion for NAS Keflavik. ***Baldur Sveinsson***

ignation, encompassed a variety of mission taskings. Besides being flown as logistical transport aircraft with VP squadrons, some units, such as VXN-8, utilized them as mission training aircraft. Several UP-3As became Advance Base Transport Aircraft. The prototype UP-3A was eventually assigned to the Naval Air Station Bermuda. Another UP-3A was eventually assigned to NAS Keflavick. Both provided regular transport of station personnel and deployed VP aircrews, as well as supplies back and forth to the remote bases.

There are other UP-3As that have been additionally modified similar to the VP-3A VIP transport Orions. These "partial VIP" transport Orions have been laid out with VIP suites, but lack most of the finer executive appointments. These aircraft were modified by local commands that they serve, often without official authorization. One served the Commander 7th Fleet-Pacific (#150504), another (#150526) for the Commander Naval Air-Pacific, and a third (# 150605) assigned to NAS Barbers point as a back up staff aircraft for the VP-3A stationed there in Hawaii.

The partial VIP utility Orions are an example of a unique UP-3A where their configurations are considerably different than the standard UP-3A and those advanced base transport aircraft. Other utility P-3As have also been utilized as airborne testbed aircraft for the research and development of advanced avionics for naval aviation. One group of R&D UP-3As operated with the Naval Air Development Center, or NADC, in Warminster, Pennsylvania.

NADC UP-3A

NADC has been around in one incarnation or another since the beginning of naval aviation. Its mission was to advance existing aircraft and airborne systems performance by exploiting future designs and technologies. It was the Navy's principle research, development, test and evaluation center for naval aircraft. It pioneered or improved upon every aircraft in the fleet and has made lasting effects on aircrews in their day to day operations; from the fire retardant flight suits and safety boots that they wear, to the performance of their

NADC UP-3A testbed Orion; "Droop Snoop" project configured. ***Georges VanBelleghem***

NADC UP-3A test aircraft; "SAR" testbed Orion. *US Navy / NADC*

NADC P-3C testbed Orion. *US Navy / NADC*

NADC "laser" Orion. *US Navy / NADC*

aircraft and the safety equipment available in the event of an emergency. NADC has had affects on the missions that aircrews fly and the way they fly them for years.

NADC operated its UP-3A and other Orion aircraft as proof-of-concept platforms testing numerous advanced systems that are now in the fleet and many still under development—some specifically designed for the P-3 Orion. It was NADC that, once having put the P-3A into service, began planning the advanced "A-NEW" program towards the P-3 Charlie model and its follow-on series of Updates. It was NADC that developed the Orion's computer software Upgrades, the IRDS infrared system, new ESM, improved MAD and acoustic systems for the P-3—culminating in one of the most advanced and electronically sophisticated aircraft flying today!

NADC UP-3A testbed Orions started out with basic configuration from an ASW P-3A to utility configuration, including the stripping out of all equipment and the instillation of slide

NADC P-3C testing ALR-77 ESM system. ***Lockheed***

Interior arrangement of NADC testbed Orions. ***US Navy / NADC***

onto floor rails to facilitate specially designed electronic racks for testing new development systems and sensors. The aircraft were also fitted with special electrical and electronic cooling outlets throughout the aircraft.

The aircraft were then modified accordingly for specific research and development projects. One NADC UP-3A (#148883—the first production P-3A) was established as a flying acoustic sensor laboratory, testing advanced next generation avionics systems. One system program tested was "Project Beartrap." Beartrap was a unique airborne intelligence data collection and processing system. It encompasses the collection of acoustic and non-acoustic data for ASW intelligence. Installed on a small number of fleet P-3s fitted through a roll-on/roll-off package, the system enhances the aircrafts' submarine detection capability and allows for development of optimum tactics , processor modes and ASW weapons capabilities supporting future fleet ASW operations.

Although Beartrap is classified, it's believed that upwards of six Orions were equipped with the system at any one time. They have been operated by regular Navy fleet patrol squadrons on both coasts. The system is centered around an additional acoustic processor and unique display sensor station located in the aircraft's starboard aft observer station. Beartrap has been continuously upgraded over the years, since its introduction in the 1970s. The most recent upgrade or next generation system is called APEX.

This NADC UP-3A was also the testbed aircraft for the Synthetic Aperture Radar project. This project consisted of a complete structural and electrical modification into the SAR testbed aircraft. The Mod included a large ventral radome mounted in the fuselage, aft of the wings. The project gathered basic SAR radar information that has led to the initiation of other Orion SAR projects.

Other R&D projects flown on this UP-3A included a joint civilian/military collision avoidance system and the High Altitude Reconnaissance Platform And Surveillance System (HARPASS) designed for other Navy aircraft.

Another NADC UP-3A (#148889) tested a variety of avionics systems, including a 360 degree look down radar unit called the "Droop Snoop," as well as a jettison capable flight data recorder. The UP-3A was also the first Orion aircraft to be equipped with the 48 externally loaded (CAD) sonobuoy chutes of the P-3C. This UP-3A also had the distinction of being the S-3 Viking Avionics System flying testbed. The Orion was re-configured inside the fuselage with the S-3 tactical sensor stations and equipped with the S-3 radar, NAV/COM, ESM, MAD and acoustic systems. Externally, the Orion sported the Viking wingtip mounted ESM antennas and belly mounted FLIR pod.

Interior arrangement of NADC testbed Orions. ***US Navy / NADC***

General Offshore Corporation UP-3A Orion. ***General Offshore Corp.***

The most unusual NADC UP-3A is the utility Orion extensively modified for laser and electro-optical systems projects. A UP-3A (#152150), known as the glass bottom Orion, was an aircraft re-engineered with four optical windows installed into structurally reinforced floors. The window bays house laser generators as part of the Laser Research project—used to detect undersea objects such as submarines. This project also gathered additional information on cloud thickness, different sea conditions, aircraft altitudes and water depth parameters that might effect the lasers' performance. Other systems tested on the aircraft include a NADC developed electro-optical surveillance systems.

It's important to mention that NADC utilized other Orion aircraft as RDT&E testbed aircraft. Most are different Update versions of the P-3C. One of these P-3C (#158204) is also fitted out for testing lasers. The "laser bird" was specially equipped with two large optical windows mounted into structurally strengthened floors. The aircraft was used for the Tactical Airborne Laser Communications Program which tested various laser systems to communicate with submerged submarines. The program was divided into two projects; the "Y" Blue/Green laser projects and the Advanced Optical Receiver project. The "Y" Blue/Green project utilized lasers of different wave lengths, demonstrating a one way, downlink of communications from an aircraft to a submerged submarine. Project AOR tested the reverse, demonstrating the aircraft's ability to receive a laser uplink from a submerged submarine. These tests are designed for the development of a satellite based submarine communications system.

Still other NADC P-3C (#158912, #160291 and #162770), having been development aircraft for the different Charlie Updates, were used for testing avionics, such as computer software, advanced acoustic processing, satellite communications and global positioning systems. These aircraft also flew other R&D projects such as the improved digital MAD and towed MAD systems. The towed MAD experiments encompassed a two hundred pound pod mounted on the outer wing station that consists of hundreds of feet of wire cable, a reel motor and a cylindrical bird with fin stabilizers. The MAD bird is reeled out, down into the ocean and towed along by the aircraft. This procedure places the MAD detector away from the interference of the aircraft and provides for more sensitive readings at greater depths.

NADC later became the Naval Air Warfare Center-Aircraft Division in 1993 and has since been re-located and absorbed by NAWC-AD Pax River in Maryland. Under the February 1995 announced NAWC upgrade program, all the NADC/NAWC Warminster UP-3As have all been re-designated NP-3D.

GOC UP-3A

Another unique UP-3A was the utility Orion modified for sonobuoy quality assurance testing by a division of General Offshore Corporation. The company was contracted by the Navy to test production sonobuoys selected from manufacturers' production before delivery to the fleet and samples taken from Navy stored stocks for effectiveness. New advanced development sonobuoys were also tested.

The unique utility P-3, loaned to GOC and flown by the company via the Navy contract, was modified with a externally loaded, multi-tube sonobuoy launching chute system installed in the port side of the weapons bay. Launching sonobuoys were also accomplished through the retained sonobuoy launching system and free fall chutes. The Buoys were launched via a control activation system mounted in the cockpit. This is a deviation from the standard UP-3A configuration established by NADEP Jax.

The GOC UP-3As were further equipped with fuselage mounted video cameras (fore and aft) to record the sonobuoy's ejection, free fall and splash down characteris-

GOC UP-3A's weapons bay mounted CAD sonobuoy chute launching system. ***P-3 Publications***

tics. Inside, a visicorder system records important airspeed, altitude, temperature and ejection velocity parameters onto paper printed images. Additional receivers and recorders tape all signals emitted from the buoys for later evaluation.

General Offshore operated out of NAS Brunswick, Maine, and forward deployed to St. Croix in the Virgin Islands where the company maintained a sonobuoy test range facility with boats also equipped with equipment for testing sonobuoys.

The quality assurance testing contract with the Navy, in force since 1965 (flown over the years by other companies and other aircraft) was canceled in late 1993 due to Navy budget cuts. The sonobuoy testing UP-3A and a companion P-3A utilized by General Offshore were returned to the Navy and subsequently retired to AMARC desert storage.

Most UP-3As still in service with the US Navy today have been improved. Under the Navy's 1994 P-3 Derivative Re-engineering Program, the UP-3As, along with Navy TP-3As and VP-3A variant Orions were re-engined with -14 engines donated by reserve (light weight) P-3Bs prior to their retirement to AMARC. This was an effort for the Navy to standardize the Orions' engine configuration throughout the fleet. The Navy UP-3As have also been re-fitted with new navaids, communications and color weather radar units.

UP-3B

In the wake of Navy Budget cuts and the retirement of most UP-3As that have reached the end of their operational service life with the Navy, a number of P-3 Bravos have been tasked to those utility missions of the retiring Alphas. Although designated "UP-3B," not all the aircraft have complete utility configuration changes associated with Orions. Some have just had their sensor equipment removed, while others have been extensively modified at foreign aeronautical rework facilities outside the United States.

EP-3J

One of the more recent P-3 variants to be developed by the Navy is the "EP-3J." The "J" model replaces older EP-3A aircraft that were configured to train carrier battle groups in electronic warfare surveillance and EW countermeasures by simulating hostile maritime patrol aircraft, anti-ship missiles and surface/sub-surface search and track radars—to sharpen the counter-attack skills of CVBG ship crews.

GOC P-3A over the Company's sonobuoy test and recovery boat based out of St.Croix, the Virgin Islands. ***General Offshore Corp.***

EP-3J C³I counter-measures Orion. ***Terry Taylor***

The EP-3J assumes the peacetime fleet EW training task and adds a new tactical command, control and communications/countermeasures (C³I/CM) capability. This capability is accomplished through the incorporation of the USQ-113 fast-scanning jamming unit. The USQ-113 is a communications intrusion/deception jamming set that is capable of scanning, monitoring and jamming multi-frequency bands of radio communications to degrade a CVBG's ability to direct their battle forces. The jammer unit also provides the EP-3J a great enough C3I/CM capability to perform a primary tactical countermeasures role if ever called upon to do so. That makes the EP-3J the Navy's only dedicated communications jamming platform—hence the "J" for jammer.

The EP-3J's mission suite encompasses the USQ-113 as its primary sensor, but also is equipped with a number of simulator pod systems. They include an APS-80 maritime search radar (simulator), the ALQ-21 third world threat (simulator), the ALQ-167 smart noise/deception jammers and ALQ-170 missile seeker (emitter) radar signal simulator, as well as the AST-4 and AST-6 threat radar signal simulators.

The EP-3Js are also provisioned to carry ALE-43 wing-tip mounted chaff dispensers. The chaff unit dispenses clouds of tiny metal strips to confuse radar systems and mask the aircraft movements and provides corridors to screen aggressor fighter aircraft.

Among the many antennas sprouting from the EP-3J are aerials for secure HF, UHF and satellite communications. The EP-3J wing hard points are equipped with universal equipment connections for quick installations of the various simulator pods. In 1996, GPS was added to the aircraft and its associated antenna.

There were two EP-3Js developed, re-engineered from ex-navy reserve P-3 Bravos (#152719 and #152745) modified by Chrysler Technologies Airborne Systems Inc. of Waco, Texas—now the Waco Division of Raytheon E-Systems—in early 1992. The first EP-3J was delivered to the Navy during March 1992 with the second following shortly there after. The Jammer Orions were originally assigned to Tactical Electronic Warfare Squadron 33 (VAQ-33) stationed at NAS Key West, Florida.

The VAQ-33 initially operated a super constellation and NKC-135 in the EW simulator fleet training role. They were later replaced by a EP-3A (#150529) in 1984 and a second EP-3A (#151368) in 1987. The EP-3J operated with the electronic warfare aggressor unit until October 1994, when Navy budget cuts forced VAQ-33 to disestablish. The EP-3J and their mission was transferred to reserve patrol squadron sixty-six (VP-66) based at

EP-3J Orion. ***Terry Taylor***

NAS Willow grove, Pennsylvania, until August 1997, when they were assigned to the newly formed VQ-11.

The EP-3J continue to be a versatile platform providing a multitude of simulated air and surface threats, thereby helping Navy combatant crews to overcome and counter threats in a hostile electronic warfare environment. The EP-3Js normally operate against the battle groups unannounced and simulate attacks during joint fleet exercise, pre-deployment workups or during carrier transits.

The EP-3J program was originally established in two phases. The first phase encompasses the current configuration with the second phase adding the ATL-40 Jammer simulators which provide directional high power jamming via ventrally mounted antennas, ALR-775 receivers, OE-320 direction finders and ESM systems for acquisition, look-through and track capabilities. Even with the capabilities current in the EP-3J, the aircraft's potential applications have not yet been totally realized.

SPECIAL PROJECTS ORIONS

You can't discuss P-3 Orion variants without examining a particular P-3 modification that to say the least is shrouded in mystery. These Orions encompass those operated by Special Projects Units. Patrol Squadron Special Projects is the designation given to two special Orion squadrons that have P-3 aircraft designed as special reconnaissance platforms utilized for surveying undersea, surface and overland targets.

These Special Projects squadrons trace their history back to the early 1950s when special detachments of regular fleet VP squadrons flew specially equipped MPA patrol planes on covert eavesdropping missions against Soviet and Communist Blocks territories. The two current Special Projects units, VPU-1 and VPU-2, were originally one crew detachments of regular VP squadrons formed in 1969, which later separated as permanent detachments with their own officers in charge,

Disguised Special Projects Bravo; bearing bureau number of a TA-4F Skyhawk Jet ***Baldur Sveinsson***

before becoming commissioned squadrons in 1982. VPU-1 developed from a detachment of VP-26 with VPU-2 being derived from a VP-4 detachment.

Each squadron operated two uniquely configured P-3 Bravos employing several unique sensors, including a special DIFAR-B acoustic system and additional electronic intelligence gathering and communications intercept systems, as well as optical surveillance systems.

Due to the highly classified nature of the Special Projects missions and to lessen suspicion of its special taskings, the original P-3 Bravos often masqueraded as regular fleet patrol squadron P-3Cs complete with borrowed squadron markings and aircraft bureau numbers painted on the fuselage to disguise true purposes. The Bravos were even decorated with pseudo-exterior sonobuoy chutes painted on the bottom of the aircraft. Occasionally these Orions were seen bearing bogus bureau numbers later discovered to be those of Navy

Special Projects P-3B disguised as a VP-10 Bravo. ***Baldur Sveinsson***

A Special Projects P-3B photograph in June 1978; the real Orion (#152724) crashed on April 26, 1978. ***Baldur Sveinssoni***

P-3s lost in flying accidents or in several cases an Israeli A-4 Skyhawk Jet. It's known that the bogus markings were changed for each flight, which added weight to the aircraft and often made the aircraft heavy and hard to takeoff. This problem was later alleviated in 1986 when all the Navy P-3 began losing their tail markings and allowing the Special Projects Orions to blend in with fleet regular aircraft at will.

It has been said that the Projects Orions are cousins of the EP-3E ARIES electronic intelligence aircraft, but employ additional unique sensors such as Laser Dazzlers and Electro-optical systems. Laser Dazzlers are laser devices that were used to counter hostile optical surveillance sensors and have been utilized by both US and Soviet forces during the Cold War—until an agreement, made during the Salt II treaties talks, banned their use except in times of war. In fact, it was a VPU Orion that was painted by a Laser Dazzler from a Soviet surface vessel in 1986, that permanently blinded several crewmen on board the aircraft.

The Projects Orions were also the first operational mission P-3s to be equipped with an Electro-optical system. Known as TOSS, short for Tactical Optical Surveillance System, the electro-optical system was developed by NADC Warminster. This electro-optical system enhanced VPU operations and permitted the aircraft to conduct overland surveillance missions during the 1991 Gulf War. Projects Orions were known to have been flown on overland surveillance missions to define the threat imposed by Iraq's air defenses in support of American and Coalition tactical strikes during the desert conflict. The VPU P-3 later provided real-time battle damage assessment (BDA) of those overland strikes and those conducted along the Kuwaiti coast.

In 1994, VPU units utilized the TOSS equipped Orions to provide non-military assistance to humanitarian (relief) aid organizations operating in Rwanda during the civil war there. The Special Projects aircraft located and monitored refugee camps in the neighboring countries.

These overland humanitarian operations later incorporated intelligence gathering flights on the warring tribes in Rwanda—due to fears that those tribes that initiated the civil war would attack returning refugees as they crossed back over the borders on their way home.

In early 1992, a program was initiated to develop replacement platforms for the two Special Projects squadrons. Four Navy P-3C Update I airframes (#159504, #160285, #160288 and #160292—which was an Update II later re-configured as the Update IV prototype aircraft) were selected as baseline aircraft for the Special Projects replacement aircraft program designated P-317. The modification was performed by the Jacksonville NADEP and standardized the VPU mission configurations among the four different bravos onto one Charlie airframe.

The interior layout of the P-317 Orions is very similar to EP-3E aircraft with sensor operator workstations positioned along both the port and starboard side of the aircraft. The new P-317 Orions are clad in the all gray tactical paint scheme with generic US Navy markings with no squadron insignia and only occasionally display bogus bureau numbers in small print under the horizontal stabilizers from time to time.

NON - MILITARY VARIANT ORIONS

Besides the US Navy and foreign operators, some of the most unique and exciting P-3 variants are operated by civilian government agencies and contractors. These derivative P-3s are involved in a variety of earth science and atmospheric research operations, as well as acting as versatile surveillance platforms employed in the war on drugs.

NASA ORIONS

The first US Government agency to acquire a P-3 Orion was the National Aeronautics and Space Administration. NASA maintains a large fleet of aircraft that has included the P-3 Orion since 1967. The first NASA P-3 was the original Lockheed YP3V-1/YP-3A aerodynamic prototype aircraft

NASA Orion. ***NASA***

Once an L-188 Electra and later the YP-3A Orion prototype, NASA acquired the aircraft and redesignated it an NP-3A remote-sensing platform; note Electra style split cockpit windows and numerous passenger windows. ***NASA***

which was re-engineered from an L-188 Electra. NASA acquired the P-3 from the Navy in 1966 and extensively modified the aircraft into a remote-sensing platform for its Texas based Johnson Space Center's "Earth Resources Program." The project involved various aircraft, satellites, the Skylab and later the Space Shuttle to collect photographic information for studying the earth's agricultural, forestry, marine and environmental resources.

The Mod included stripping out the remnants of the P-3 interior and any left over equipment racks, components, workstations or systems of the past Navy P-3 development program. The aircraft was then modified with floor tracks throughout and a permanent navigator's station installed behind the cockpit on the starboard side. All other mission operator stations were of a roll-on/roll-off configuration, as were all the mission sensors and electronic equipment. Universal equipment racks accommodated temporary project stations, mission equipment and additional seating, giving the interior an entirely new look.

Additionally, the aircraft was configured with two instrument bays, one unpressurized in the area of the P-3's weapons bay, forward, and a pressurized bay in the aft section of the aircraft. The aircraft was further equipped with several air conditioning units to cool the various mission electronics and laser units. Mission systems encompass internal and Omega navigational units, GPS, azimuth radiometers, lasers and a number of video and photographic cameras for numerous research mission taskings. The aircraft was permanently equipped with a primus 400 weather radar.

Upon completion of the Mod, the aircraft was re-designated NP-3A. The "N" designation is that which is given to an (Orion) aircraft that has been so drastically modified from its original mission configuration that it's beyond any practical economic limits to be restored to the aircraft's former use.

Although the outward appearance of the NP-3A is that of an Orion, the heart of the aircraft is an L-188 Electra and retains several Electra characteristics; such as a roll-up main cabin door and split cockpit side windows, as well as numerous passenger windows. As the Orion aerodynamic prototype, the aircraft was modified with P-3 type wings , which later suffered severe corrosion with NASA and were replaced with Electra wings. The NP-3A's engines were Allison 501D turboprops common to the Electra with one-of-a-kind Hamilton Standard propellers. In essence, the NASA NP-3A was a very unique Orion aircraft.

In September 1977, the NP-3A was transferred to the NASA flight facility at Wallops Island, Virginia. Prior to its retirement in February 1993 and subsequent transfer to the National Museum of Naval Aviation in Pensacola, Florida, the NP-3A took part in many science related missions. Although most of its missions were space related, others involved ocean

NASA's new "EFIS" P-3B Orion remote-sensing / airborne research aircraft. ***NASA***

NASA "EFIS" P-3 Bravo. ***NASA***

physics, meteorology, atmospheric chemistry and earth science studies for NASA and user scientists worldwide. Some of the studies included outfitting the Orion with unique sensors and consisted of projects such as the Global Ocean Flux Study—to measure the bio-geochemical ability and rate at which the world's oceans store excess carbon dioxide from the atmosphere through a survey of Phytoplankton distribution. Other projects included SEASAT Imaging, Radar-B and Labrador Extreme Wave Experiments where the NP-3A conducted radar and laser wave elevation profiles as a preliminary study for the development of the Maritime Dangerous Wave satellite warning system. Another project, the Stratosphere Aerosol Measurement Project, studied high concentrations of particles and gasses suspended in the stratosphere, like those from volcanic eruptions that affect climate and weather world wide.

NASA's second P-3 Orion was received during September 1990 from the US Naval Reserve. This Orion, a P-3B (#152735), was modified for the NASA airborne research mission with most of the remote-sensing equipment and capabilities inherent in the NP-3A. Additionally, the NASA P-3 Bravo was upgraded with an improved avionics package highlighted by an advanced Electronic Flight Instrumentation System, or "GLASS COCKPIT." Installed by Associated Air Center in Dallas, Texas, the modification encompasses the Honeywell EDZ-805 Glass Cockpit systems with five data displays replacing most of the older flight instruments, a new radio rack including SATCOM and an upgraded navigation

The Honeywell EDZ-805 electronic flight instrumentation system installed onboard the NASA P-3B Orion. ***NASA***

Department of Commerce / NOAA WP-3D Orion. ***NOAA***

suite incorporating dual GPS units. The Aircraft is also equipped with a new ground proximity-cockpit warning system, a cockpit voice recorder, emergency locator transmitter and advanced autopilot systems, not yet available on the newest passenger airliners.

The NASA P-3B operations are further enhanced by the aircraft's larger fuel capacity and Auxiliary Power Unit. The APU provides power to run experiments and equipment on the ground in remote locations world wide. The EFIS Bravo's first mission encompassed operating out of Alaska and Greenland as part of a joint North Pole Area Survey Project with the US Navy.

The EFIS Bravo was joined by another reserve P-3B (#153429) at Wallops Island in 1993. This second Bravo was acquired as an operational mission aircraft to replace a retiring NASA L-188 Electra remote-sensing platform. But due to unexpected NASA budget cuts, the aircraft was later transferred to AMARC for desert storage as a parts locker.

At NASA's Wallops Island facility Orions continue world wide scientific surveys and range support to the Space Shuttle program—in the future, the P-3 will be the remote-sensing platform at the forefront of those operations.

NOAA WP-3D; note exterior features of the variant Orion aircraft. *Lockheed*

NOAA ORIONS

One of the most unusual P-3s in the Orion Family of aircraft is a unique variant model of the P-3C, designated "WP-3D." The WP-3D is a flying laboratory for airborne atmospheric and oceanographic research. The aircraft was a production derivative of the P-3C, especially built by Lockheed for the Department Of Commerce and operated by the National Oceanic and Atmospheric Administration, or NOAA.

Since 1975, the WP-3Ds have been the principal aircraft utilized for hurricane research, penetrating hundreds of tropical storms into the interior eye to gather vital data needed to understand how these tropical cyclones form and detect any characteristics that can assist in predicting possible tracks. The tracks are important to the National Hurricane Center for developing long-lead 24 and 36 hour forecast models. The greater the accuracy of these models, the more time coastal regions in the hurricane's path have to evacuate and the loss of life is decreased.

There were two WP-3Ds, N42RF (#159773) and N43RF (#159875), that rolled off the Lockheed Charlie production line in June 1975 and January 1976 respectively. They were delivered in a neutral state, devoid of interior stations and electronic racks. The airframes were designed and built with structurally strengthened floors and tracks running from behind the cockpit aft all the way back to the main cabin door. The multi-functional sensor stations and equipment racks onboard the aircraft today are merely roll-on/roll-off units designed by NOAA for various mission profiles. Although the hurricane research configuration can be considered the most

Close-up of WP-3D candy-cane striped wind gust probe. *P-3 Publications*

WP-3D Orions are multi-mission aircraft and perform a range of taskings such as fisheries- law enforcement. *NOAA*

common, there is no permanent or standard layout. The WP-3D is by design a flexible remote sensing platform.

Besides its unique paint scheme, the WP-3D is denoted by a twelve foot diameter ventral mounted radar radome that houses a powerful 360 degree looking C-Band radar. This radar provides horizontal profiles of storms, while the large fat tail radome houses a side-looking Doppler radar for clean-cut, vertical views. Lastly, the aircraft possess a candy-cane striped "wind gust probe" protruding out of the nose, somewhat reminiscence of a Narwhal.

The WP-3Ds were also the first production P-3s with a cabin rating over thirty thousand feet and the first P-3 with a commercial color weather radar, as well as the first civilian P-3 with Harpoon missile system wiring in its wings used for wing mounted sensor pods.

The WP-3Ds were the first weather research/reconnaissance aircraft with the Aircraft Satellite Data Link system, or ASDL. This data link system provides for the transmission of digital radar and other meteorological data to the Hurricane Center in real-time via satellite. Previously, reconnaissance aircraft transmitted observations and numerical data by radio and voice links.

The WP-3D are operated and maintained by the NOAA's Aircraft Operations Center currently residing at MacDill AFB in Tampa, Florida. The AOC moved from its previous operating base, the Miami International Airport, in January 1993 where NOAA hurricane research aircraft were flown for over 38 years. The WP-3D replaced venerable old DC-6 planes that participated in hurricane seasons up until 1976, when congressional money, appropriated to modernize the NOAA research fleet and to upgrade instrumentation, spawned the development of the new Orion "Hurricane Hunters."

The NOAA Orions are by design airborne ocean-atmospheric research platforms that are utilized for more than just hurricane missions. The two WP-3Ds have participated in numerous research projects over the years that have included the Joint Airport Weather Study (1982), Oklahoma-Kansas Per-Storm Project and Equatorial Pacific Ocean Climate Studies (in 1984), Project Ocean Storm (1987), the GULFMEX Investigation and Fisheries Oceanographic Combined Investigation (1988), to name but a few.

The aircraft were even once utilized in a law-enforcement mission in conjunction with NOAA's National Marine Fisheries Service. Operating out of Seattle, Washington, a WP-3D flew ocean surveillance flights to locate suspected illegal foreign fishing vessels poaching in restricted Alaskan waters.

Other non-hurricane related research capabilities of the WP-3D have included participation in projects such as the Tropical Ocean Global Atmosphere program and its integral

NOAA WP-3D forging towards the future. *NOAA*

Couple Ocean Atmosphere Response Experiment, or "TOGA-COARE." In these ocean-atmospheric interaction experiments, the NOAA P-3s played a vital role as data collection platforms. The "meteorological AWACS," as the aircraft were referred to during the project, operated at low altitudes between one hundred and ten thousand feet sampling assigned areas gathering temperature and moisture fluxes as well as profiling cloud dynamics. The NOAA Orions were specially modified with a C-BAN Scatatrometer for the TOGA-COARE project to measure sea surface winds.

Another recent project that the NOAA P-3s have been taking part in is the Southern Ozone Study, or SOS. The SOS project involves the continuous collection of regional Ozone concentration measurements and weather/climate related factors throughout southern areas of the United States. The WP-3D act as air quality research aircraft tasked with collecting detailed, regional scale air chemistry and meteorological measurements throughout eight southeastern states.

The SOS goal is to design and implement scientific research and modeling programs to promote scientific and public understanding of excessive and harmful levels of Ozone air pollution (or SMOG) accumulating in the near-ground atmosphere.

Similar to the hurricane research mission, WP-3Ds also provide airborne support to the annual VORTEX projects. Initiated by the NOAA's National Severe Storms Laboratory in Norman, Oklahoma, the WP-3Ds collect research data on tornadoes that quickly form throughout the Midwest during the spring and early summer months each year. The Orions utilize their Doppler radars during and after tornado formation to profile the supercell thunderstorm dynamics. Besides utilizing their radar, the NOAA WP-3Ds also employ the aircraft's side-looking video camera system to record visual characteristics as they fly along the dry- line boundary areas of super cell thunder storms prior to their formation.

The future of the WP-3D will continue the same as the last twenty years, but with expanding utilization. Besides seasonal hurricane research and severe storm studies, the WP-3D will continue to be involved in projects with shifting emphasis toward broad based environmental studies. These include adaptive cooperative atmosphere and air chemistry programs and working with other research institutions and universities.

There are plans to modernize many of the instruments onboard the aircraft and upgrade the systems to standardize the output formats with those of the rest of the research community. There are also plans to provision the WP-3Ds to carry the "AMPS" pods and its associated sensor operator station as a means to provide additional operational test flight capabilities to the AMPS program. In the meantime, the WP-3Ds continue to be the versatile meteorological research platforms that they are and will remain so far into the next century.

CUSTOMS ORIONS

In an effort to stem the flow of illegal drugs entering the United States, the US Customs Service adopted a unique P-3 variant in 1987 to provide long-range detection of aircraft and surface vessels suspected of smuggling contraband narcotics. These new derivative Orions are designated P-3 Airborne Early Warning and Control Platforms, or "P-3 AEW&C."

The P-3 AEW&C Orions stem from a mid-1980s decision by Lockheed to develop a P-3 Orion and C-130 Hercules versions of AEW aircraft to compete with the established E-2C Hawkeye and E-3A Sentry early warning and control aircraft, with potential customers (at the time) including Australia, Japan and the United Kingdom. Lockheed subsequently produced the AEW version of the P-3 Orion. The company took one of the ex-Royal Australian Air Force P-3B airframes (#155299), traded in towards the purchase of newer Charlie Orions, and converted it with a twenty-four foot diameter Raytheon APA-171 rotodome. This AEW aerodynamic prototype, civilian registered N91LC, first flew on June 14, 1984, devoid of any mission systems or equipment. The prototype aircraft was later flown on April 8, 1988 as the P-3 AEW&C mission systems demonstrator aircraft complete with APS-125 radar system and other C3 mission equipment. The mission equipped prototype was established to proof-of-concept trial the AEW&C mission suite and demonstrate its potential to customers. Besides the unique dorsal mounted rotodome and multi-color paint scheme, the AEW&C variant Orion was the first P-3 to be modified with a glass cockpit. The unique feature encompassed the removal of all standard electro-mechanical flight instruments and installation of the Collins EFIS-86B electronic flight instrument system comprised of

US Customs Service P-3 AEW&C Orion. ***Lockheed***

Lockheed P-3 AEW&C prototype aircraft. ***Lockheed***

US Customs first P-3 AEW&C; note original paint scheme. ***Lockheed***

five 5x6 inch multi-functional color displays. The EFIS screens display standard flight data, as well as additional color weather radar and navigational information. Self-diagnostic checks and preflight check lists can also be displayed through the system.

Two color displays are positioned for each the pilot and co-pilot with the fifth display centered between them for shared flight information or displaying the color weather radar data from the TWR-850 Doppler weather radar system.

Although designed for military applications, the only customer that became interested in the AEW&C Orion was the US Customs Service. The capabilities of the P-3 AEW&C were ideally suited to the US Customs counter-narcotics mission to detect, classify and direct interceptions of suspected aircraft along the southern-most borders of the United States and out into the Gulf of Mexico and the Caribbean.

The US Customs service, under the Department of Treasury, officially placed an order for the first P-3 AEW&C N145CS (#155299) on June 12, 1988 with an option for two more aircraft in the future. Customs took delivery of the first "Dome," as the aircrews would later refer to them, on June 17, 1988. Shortly thereafter, a second Dome, N146CS (#154605), was ordered and subsequently delivered during April 1989.

P-3 AEW&C

The US Customs P-3 AEW&C Orion's primary sensor is the General Electric APS-138 radar with a detection coverage ranging from sea level to a height of one hundred thousand feet, out to approximately 200 nautical miles. The APS-138 differs from the APS-125 radar system, originally installed on the prototype and delivered with the first aircraft, through the incorporation of the Advanced Radar Processing Sub-system or ARPS. The ARPS system increases the sensitivity of the radar while minimizing noise, clutter and the false alarm rate of the unit. It provides the APS-138 with a detection capability over both land and sea and can display over 2,000

USCS P-3 AEW&C in all-gray paint scheme. ***Terry Taylor***

The third Customs P-3 AEW&C. ***Lockheed***

The fourth P-3 AEW&C for US Customs. *Lockheed*

Customs P-3 AEW&C tactical compartment; miligraphic displays. *Lockheed*

Close-up of AEW system display. *Lockheed*

contacts simultaneously. The ARPS comprises a coherent UHF Doppler radar with a high-track resolution and airborne moving-target indicator.

The aircraft's twenty-four foot rotodome has greater rotodome-to-fuselage separation, which provides almost twice the look-down angle (over the wings) than the same dome on the E-2 Hawkeye. Built into the rotodome is a sixteen foot wide identification friend or foe (IFF) interrogator. The IFF system antenna is an integrated identification (I-Band) high gain array which can track up to twelve hundred contacts per revolution.

The aircraft is also equipped with a passive electronic support measures (ESM) unit that provides positive emitter identification and augments the radar to give a complete real-time tactical picture of airborne and surface activity. The combined systems create an anti-jamming and interference-free capability, allowing for operations in a hostile electronic environment.

The radar and its associated mission sub-systems are integrated through an onboard AYK-14 digital processor. Raw radar data is fed into the computer and processed with the additional information generated and transferred to the operator's station by a digital databus. Raw radar data is also simultaneously fed into radar video scan converters that produce various video outputs. These video signals are also routed to the operator's display and appear as a radar track history overlays on the tactical plots.

The P-3 AEW&C's radar operator's displays are Saunders Miligraphics nineteen inch color units with touch sensitive "fingers-on-screen" controls. The miligraphic displays also encompass military graphic symbols, color-coded according to the level of importance and a built-in data processor for the touch screen controls. The system also uses standard trackball and keyboard controls. The Customs AEW&C also

The P-3 AEW&C's glass-cockpit. ***Lockheed***

The first Customs P-3A Orion. ***US Customs***

retains the aircraft's original APS-80 radar for additional sea search and detection.

The Customs P-3 AEW&C encompasses an extensive communications suite consisting of five UHF/VHF radios and an air-to-air, air-to-ground datalink system. The main communications system is a SATCOM unit that links all onboard communications to ground command, control, communications and intelligence (C3I) centers which are jointly operated by USCS and the US Coast Guard at centers in Miami, Florida, and Riverside, California.

The P-3 AEW&C mission initially provided radar coverage for the approaches to the US southern borders, from the coastal waters of California and Baja through the Gulf of Mexico and into the turquoise waters of the Caribbean.

The Domes were actually acquired to augment the string of "AEROSTAT" radar balloons that are stationed along the southern perimeter of the US and those dotting the Caribbean. When down for maintenance or when there was adverse weather conditions, the Dome Orions were designed to fill the gap. But by the end of 1989, as soon as the first Dome made its appearance, narcotics traffickers switched tactics. Smugglers began looking elsewhere to find new routes into the United States. This prompted a new policy to step-up air interdiction activities and take the war to the smugglers. This move relied heavily on the P-3 AEW&Cs and their sophisticated detection and communications capabilities. The new policy included operating from Custom's Surveillance Center at NAS Corpus Christi, Texas, and deploying the Customs aircraft to military bases in Puerto Rico and Panama to detect suspect planes leaving Columbia and other known trafficking nations to trans-shipment points in Latin America and the Caribbean.

US Customs Service P-3A "SLICK" Orion. ***Terry Taylor***

With the successes of the new Customs interdiction policy, demonstrating the importance of the P-3 AEW&C, funding for a third Dome was approved and a fourth proposed P-3 AEW&C began to gain acceptance within the Treasury Department. The third Dome, N147CS (#152722), was delivered in July 1992, with the fourth, N148CS (#154575), following by late 1993. Both were developed from ex-USN P-3 Bravo Baseline airframes.

P-3A SLICKS

The P-3 AEW&Cs were not the first nor the only Orions that the US Customs Service flies. Customs also operates four US Navy P-3A Orions that it utilizes for long-range airborne interceptions. Although designated UP-3As by the US Navy, the Customs service considers them as P-3A Orions and has nicknamed them "SLICKS" in contrast to the Domed Orions.

The Slicks are equipped with a nose-mounted APG-63 multi-mode radar (the same fire-control radar installed aboard the F-15 Eagle), an AAS-36 IRDS infrared detection set and an advanced communications suite including a SATCOM and other data link systems. They conduct long-range intercept and communications surveillance flights often working in tandem with Customs P-3 AEW&Cs as "Hunter-Killer" teams to detect, track and then intercept suspected smuggling aircraft.

Vectored in by the Domes, the Slicks intercept the suspect aircraft and position themselves behind the plane for positive identifications. They then follow the target aircraft to its landing point or drop zone. The Slicks can then video record

USCS P-3A Slick Orion in original markings. ***US Customs***

an off-loading or air drops of contraband and direct ground law enforcement authorities to the location to apprehend and prosecute the smugglers.

Although principally modified the same for the interdiction mission by E-Systems, the four P-3A Slicks are additionally equipped for special taskings. Two are outfitted for long-range interceptions with a third possessing an additional communications relay and radio frequency intercept surveillance capability. This aircraft was instrumental in the hunt for the legendary Columbian drug cartel leader Pablo Escabar. The Customs Slick flew over the Colombian city of Calli and searched for the sounds of Pablo Escabar's voice on cellular phone frequencies. The aircraft eventually pinpointed his general area, leading Columbian Police to his location and his subsequent downfall.

The fourth Slick is a highly classified special mission aircraft and an airborne testbed for new development systems to be trialed for Customs interdiction applications. These include the recent incorporation of ASARS radar systems. The Advanced Synthetic Aperture Radar System installed on the Customs Slick is a SAR radar once utilized by SR-71 reconnaissance aircraft. The ASARS provides Customs Slicks with a long-range, stand-off reconnaissance capability that permits overland intelligence gathering flights with the highest degree of safety and the ability to not violate the airspaces of Latin American sovereignties.

The fourth Customs Slick has also been equipped with the Customs Airborne Stand-off Surveillance Systems. The CASOSS is a production AVX-1 Cluster Ranger electro-optical system employed in the interdiction mission. It provides the aircraft with the ability to verify aircraft identification, reducing the hazards of close in flying with suspected drug smugglers. The electro-optical system also supports long-range surface detection and overland surveillance.

After borrowing several Navy Orions for mission trials in the early 1980s and having seen the potential of the aircraft, the US Customs requested its first P-3A in 1983. The first aircraft (#150514) on loan from the Navy was modified by E-Systems (now Raytheon E-Systems) for the counter-drug role. Three additional P-3 aircraft (#151390, 151395 and 152170) were quickly requested and received.

The Slicks' interior layout incorporates the P-3A/B tactical workstation bench with a communications management system, IRDS display and controls and the APG-63 radar display installed next to the APS-80 radar position. There is an additional APG-63 display located in the cockpit.

The Customs P-3A Slicks were further enhanced in 1990-91, receiving standard depot level maintenance (SDLM) rework. Performed by NADEP Alameda, the aircraft were re-engined with -14 engines and painted all-over tactical gray. The new paint job corresponds to the TSP paint scheme being applied to the new P-3 Domes.

Like ghosts in a fog, the US Customs Service "Hunter-Killer" team; a Dome and Slick on the flight line. ***P-3 Publications***

P-3 "AEROSTAR" air tanker with Aero Union Corporation. ***Aero Union***

As of may 1996, the four Customs Slicks on loan from the Navy were officially transferred to US Customs Service and are now owned by the law enforcement Agency.

P-3A AEROSTARS

In the mid 1980s, World War Two era piston engine aircraft utilized for the business of airborne forest fire fighting were fast exceeding safety limits in operational service. The whole US Forest Service contracted fleet of fire fighting aircraft were in serious risk of being grounded and it prompted the immediate acquisition of new aircraft.

In 1988, the US Navy received a request from the US Forest Service for surplus turboprop aircraft to be subsequently converted into fire fighting air tankers. By 1989, the Navy made available approximately one dozen surplus ASW P-3A Orions to the Forest Service and their contractor operators.

Although this arrangement to acquire the modern turboprop aircraft would later come under suspicion of improprieties and misrepresentations within the Forest Service, the fact remains that twelve P-3A Orions were distributed between three Forest Service contractors. Of the three, one would go on and develop a new fire fighting derivative of the venerable P-3 Orion. That company, Aero Union Corporation of Chico, California, was the first contractor to receive P-3 Orions. It was Aero Union that determined the P-3 Orion was superbly suited for the role of a fire fighting air tanker.

The P-3 Orions have many inherited flying qualities that are well suited to the requirements of air tankers and the forestry mission—a sturdy, heavy-duty airframe, exceptional handling qualities and good short-field performance to name but a few. The Orion's high-speed cruising and low altitude, slow speed maneuvering capabilities permit the aircraft to dash to a drop zone and sink into narrow valleys where older,

Aero Union P-3A Air tanker. *Aero Union*

more sluggish aircraft have been excluded. The P-3 Air Tankers have more than sufficient power to operate out of remote civilian airports even with a load.

Aero Union selected the best of the subsequent nine P-3A Orions it would acquire and completely stripped them out. All unnecessary weight, including electronic racks, work stations, air conditioning components and the aircraft's galley and cabinets were removed. Even the aircraft's interior wall coverings, insulation and non-essential wiring were stripped out, leaving a large empty fuselage down to the ribs. The Aero Union Air Tanker Mod begins with the incorporation of a three thousand gallon tank that is designed, manufactured and installed by the company. These eight-door, high rate of discharge controllable tanks can dump as much as three thousand gallons of "PHOS-CHEK"—a fire retardant chemical powder mixed with water. Unlike water drops to extinguish the flames, the PHOS-CHEK slurry clings to the trees and other vegetation, forming a barrier to contain and snuff out fires. The pilot simply selects the number of doors required to produce the desired coverage.

The P-3A "AEROSTARS," as the Orion Air Tankers came to be designated, were originally scheduled to have their two outboard engines and nacelles removed, but the engines were retained and the added power provides for near vertical climbs out of air drops into very tight valleys.

Since 1990, P-3A Aerostar tankers have responded to countless forest fires in state and national forests and parks throughout California and the southwestern states. The P-3 Air Tankers have also been demonstrating their capabilities to several foreign nations, with France willing to contract for the aircrafts' services. Greece at one time was interested in acquiring some surplus P-3A airframes and planned to have Aero Union tank them up. There was even a proposed plan for Aero Union to acquire additional aircraft for a Seattle company and convert them into air tankers to fight oil well fires in Saudi Arabia and Kuwait.

Of the original nine Orion aircraft acquired by Aero Union , #152359-Tanker 24, #151361-Tanker 25, #151369-Tanker 27, #151372-Tanker 23, #151385-Tanker 21, #151387-Tanker 22, #151391-Tanker 00, have been in service with the company. Tanker 24 (#151359) crashed into a Montana mountain on October 6, 1991. The two remaining aircraft, #151355 and 151377 were later taken back by the US Forest Service and along with Tanker 21 were turned over to the Government Services Administration to be auctioned off. They were subsequently purchased back by Aero Union from GSA in October 1996.

Orion air tanker. *M. Peltzer*

Orion air tankers on the flight line ; note the tank-dumping system. ***Aero Union***

Aero Union was not the only Forest Service contractor to acquire Orion airframes. Hawkings and Powers Aviation of Graybull, Wyoming, received two P-3A airframes (#150510 and #150529) with Black Hills Aviation (now Neptune, Inc.) of Almagordo, New Mexico, collecting one P-3 (#150513). These Orions have also been recently taken back by the US Forest Service and are to be auctioned off by GSA sometime in the future.

UNACCOUNTABLE ORIONS

In the context of this publication, it must be acknowledged that there exists another group of P-3 Orions that operate in what aviation enthusiasts call the "Black World of Covert Aircraft." This group of Orions that are about to be described do not appear on any Navy service bureau lists or government records, nor are they listed on any Lockheed production files. But they were built by Lockheed nonetheless!

This is actually quite a common practice by aircraft manufacturers with US military contracts. They produce an additional number of aircraft that instead of turning the usual left off the production line, turn right and disappear into the black world of covert airborne reconnaissance and surveillance.

In regards to these P-3 Orions, there is believed to have been between twenty and twenty-four P-3 Alpha/Bravo extra production airframes produced by Lockheed during the mid 1960s at the height of Orion production activities. These extra P-3s were not assigned Navy bureau numbers nor Lockheed serial numbers and officially do not exist. These Unaccountable Orion airframes went on to receive various unique modifications believed to encompass clandestine reconnaissance roles throughout the world.

One aircraft currently flying is unofficially known as "Fat Cheeks" because of the large bulges on either side of its fuselage just aft of the flight station. It's also believed that this aircraft replaced another Unaccountable Orion similarly modified and which operated for a number of years under the supposed moniker of "Chipmunk."

Other Unaccountable Orions have less obtrusive exterior modifications, but still contain a variety of special mission systems for unknown mission taskings. These aircraft were usually dressed in generic Navy markings to blend in with those Navy P-3s operating worldwide. For anyone that works within the P-3 community and basically knows where all the known Orions are supposed to be in the world (and is familiar with their operations), the Unaccountables tend to stick out. Aircraft stored here and there, where there aren't supposed to be any or a couple P-3 aircraft flying around in black paint schemes twenty-seven years after the only known black P-3s (used by the CIA) have long been extinct also stand out. There is at least one P-3 disguised as a generic Navy fleet Orion that upon closer examination is found to be considerably longer than a standard Orion—almost as long as the L-

Unaccountable P-3 Orion. ***Tourville via Isham collection***

188 Electra that the P-3 Orion was derived from. There have been a number of stories of these ghost aircraft lost over hostile territories on covert operations. There is even the possibility that some of these "Unaccountable" aircraft were secretly assuming the identities of known P-3s that were lost on unknown missions due to hostile fire.

Another rumor indicates that at least two of the twenty-four Unaccountable Orions were constructed from spare parts and the remnants of crashed Navy P-3 airframes. In any case, the fact remains that these Unaccountable P-3 Orion aircraft exist and that they constitute a unique set of P-3 Variants in the scheme of Orion History.

In conclusion to the detailing of various P-3 variant aircraft in the world, it must be noted that the proceedings are by no means all the P-3 variants that exist, but just those developed by the US Navy for special purpose taskings and those for government agencies and civilian contractors. There are still a number of variants specific to foreign operators under development with many more to be developed in the future as ORION operations move into the 21st Century.

Chapter 6
Orion - The Next Generation

It has been thirty-five years since the P-3 entered service with the US Navy—thirty-eight years since development began—and although production has been modest, the Orion has seen an illustrious and exciting history. Now, as the future looms on the horizon, the P-3 continues to experience ever growing changes, enhancements and planned improvements.

The most interesting potential P-3 program affecting the future of the aircraft is the establishment of a new, next generation "Orion 2000" variant. The Orion 2000 was based on requirements issued by the United Kingdom's Ministry of Defense to procure a MPA replacement aircraft for its fleet of Nimrod MR2s.

The "RMPA" Program, short for Replacement Maritime Patrol Aircraft, sought to replace the British Nimrod that was fast nearing the end of its useful service life, to overcome the operational limitations and the increased cost of ownership of both an aging airframe and elderly mission system by the turn of the century. The UK needed a new aircraft in which to conduct maritime surveillance of territorial waters and sea routes, as well as providing ASW and ASUW protection. Additional requirements also call for the aircraft to possess ELINT and electronic countermeasures capabilities. In effect, the UK RMPA aircraft was to have the capabilities of the older Nimrods, plus those of a modern multi-mission, versatile maritime patrol aircraft.

The RMPA program evolved into a competition between several of the top aerospace manufacturing giants. They consisted of a British Aerospace/Boeing proposal to refurbish the Nimrod airframe with new advanced avionics, sensors and weapons capabilities. Another contender (who dropped out early) was Dassault Aerospace with a proposed Atlantique III equipped with advanced Allison AE2100 turboprop engines. One unusual offer posed by Loral-Asics/E-Systems (Now Lockheed Martin Federal Systems) was the "P-3 Valkyre" which is based on ex-US Navy P-3A/B Orion airframes upgraded with new mission systems and new Rolls-Royce Allison AE2100 turboprop engines. The last proposal was Lockheed Martin (Georgia)/GEC Marconi's Orion 2000.

The Orion 2000 next generation P-3 comprises a new technology production airframe utilizing improved materials and manufacturing techniques with great emphasis on corrosion protection of predictable areas. The twenty-five plus year airframe boasts an increased payload to lift new loads and weapons, improved fuel efficiency (equating greater ranges) and an increase in overall reliability and maintainability over the more recently built P-3C Update III aircraft.

The new Orion features modernized, advanced mission avionics through modular processing based on a high wide band fiberoptic (digital) data network. The open architecture accommodates customized mission systems and sensors

P-3C Update III+ Orion. ***Baldur Sveinsson***

P-3 Valkyre. ***Lockheed Martin Federal Systems***

Orion 2000. ***Lockheed***

tailored to the customer. The proposed Orion 2000 tactical compartment is designed very similar to the Canadian CP-140 Aurora, in a "U" shaped configuration—positioned over the wing on the port side of the aircraft. The acoustic station facing aft on the one side, the non-acoustic sensor stations on the other and the middle operator stations reserved for the TACCO, and NAVCOM. The configuration includes a spare sensor position to be utilized as a sensor manager or a duplicate sensor station.

All the new sensor operator-tactical workstations were equipped with universal, multi-functional, high-resolution color displays that can be quickly re-configured as changing mission priorities demand. An eighth sensor station is installed in the Orion 2000 aft of the main cabin door on the port side. This workstation is designed for trials of new and emerging mission systems or for the installation of additional current stand-alone sensors such as an electro-optical system.

The proposed Orion 2000 encompasses a re-configurable two-man flight station with automated digital controls and electronic flight instrumentation system (EFIS) Glass-Cockpit. Borrowed from the Lockheed Martin's C-130J Hercules, the space age flight deck promotes improved situational awareness, greater safety and reliability. It reduces the pilot's workload and includes an advanced flight management system, auto crew warning and alert system and an flight control system with auto-throttle and a INS/GPS navigation system.

The aircraft possesses a new propulsion system based on the advanced Roll-Royce Allison AE2100-D1 turboprop engines with six-bladed Dowty R411 modular composite propellers. The 6,000 hp engines and composite propellers again are borrowed from the C-130J.

The new Orion also includes a new sonobuoy delivery system, as well as all the new MPA weapons and survivability capabilities encompassed in the post P-3C AIP configuration.

The RMPA Orion 2000 additionally features options including a in-flight refueling probe, an improved ESM systems capability with near ELINT characteristics and "defensive aids sub-systems." This DASS package comprises a radar warning receiver, missile warning receiver and laser warning receiver for threat detection with chaff/flare countermeasures dispensing, as well as a proposed towed radar decoy. RMPA requirements also identify the need for an electro-optical system with long-range identification and stand-off capabilities. The RAF has tentatively described the sensor as an electro-optical search and detection system, or EOSDS.

Beginning in 1995, the RMPA program proceeded with a year of proposals and evaluations. The UK's Minister of Defense finally issued a response to the RMPA competition in late 1996. The UK selected the British Aerospace/Boeing proposal to refurbish the Nimrod. The Nimrod 2000, as it was called, includes refurbishing existing Nimrod MR2 airframes and re-equipping the aircraft with new technology avionics to give them improved capabilities and those systems already described above

Despite this eventuality, the Orion 2000 continues to be offered by Lockheed Martin with various potential sales throughout the world, including the US Navy. The future Navy maritime patrol aircraft must be based on a multi-mission platform equipped with the latest state-of-the-art systems and sensors. It must be a platform capable of reconnaissance and surveillance capabilities and the ability to carry out a complete combat mission from the search, to detection and on to the attack whether with its own weapons or in concert with other strike elements over the horizon. The P-3 Orion 2000 fulfills and exceeds those requirements.

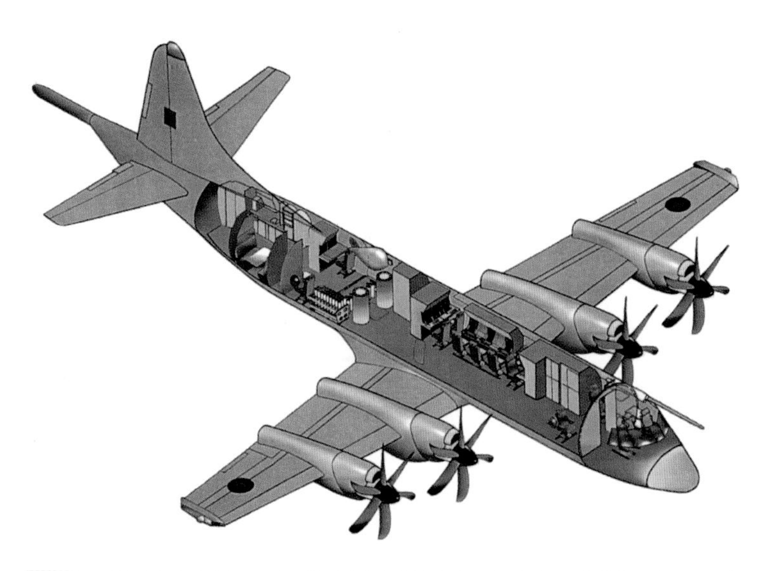

Orion 2000 interior arrangement. ***Lockheed***

P-3C Update III ***Baldur Sveinsson***

NEW POTENTIAL ORION OPERATORS

With potential sales for P-3 Orions throughout the world, several nations have been following the outcome of the RMPA program with keen interest. Germany has continued to evaluate the need for a replacement maritime patrol aircraft since it had been the lead-off customer and proposed co-developer of the P-7A back in the late 1980s. It has postponed a decision a number of times, often opting to upgrading their fleet of eighteen BR1150 Atlantic (I) MPA patrol planes in the meantime.

The Orion 2000 or any follow-on P-3 could provide the German Navy with new capabilities to fulfill its MPA requirements far into the next century, as well as providing commonality of operations and logistics compatibility between its neighbors Norway and the Netherlands.

Other potential future customers for the Orion 2000 include Italy and Saudi Arabia. Italy, another refugee from the proposed P-7A program, is also studying the need for new maritime patrol aircraft to replace its aging fleet of BR1150 Atlantic (I) patrol planes.

Since the Gulf War, a need for a domestic Gulf-based maritime patrol capability has existed. The P-3, having demonstrated its capabilities so admirably during the Gulf Conflict, has acquired the attention of several Persian Gulf Principalities. Initially, it was suggested that a Persian Gulf multinational maritime patrol unit be established. The multi-national force, led by Saudi Arabia and made up of personnel from Bahrain, Kuwait, Oman, Qatar and UAE (United Arab Emirates) would consist of ten to twelve P-3C Orions. Such a force would greatly extend the Gulf maritime picture. Separately, several of these nations are considering purchasing their own Orions—UAE looking to acquire four to six aircraft and Oman interested in two to three Orions. Saudi Arabia has also previously expressed an interest in six to eight new production P-3s separately. This is part of an ongoing Saudi military modernization program in the wake of the Gulf War, that encompasses fighter aircraft , airborne tankers, and early warning and ground attack aircraft. The cost and availability of new Orions have been concerns, but consideration for the P-3 continues. The US Navy has even suggested providing Saudi Arabia with TACNAVMOD P-3 Bravos (in US desert storage) as an interim measure to get them into an Orion program. This would establish pilot training and the initiation of an Orion logistics program. As of late 1996, no decision has been made either way.

Other possible new Orion operators span the globe. In the Pacific, Korea continues to establish funding for at least another eight P-3C Update III Orions for a new P-3 squadron to be located at Cheju-do. There is also some talk of South Korea acquiring a number of used ex-navy P-3A/B airframes as pilot training aircraft.

Like Korea, Taiwan is situated close to potentially hostile neighbors, one of which has acquired Russian KILO-class

Older P-3 airframes ; much like those being considered by Korea as pilot training aircraft. ***Lockheed***

Older P-3 Alpha and Bravo airframes like this could potentially end up in Latin American air forces. ***via Baldur Sveinsson***

diesel-electric submarines and a license to build more submarines in the future. With recent emphasis placed on acquiring ASW capabilities, a purchase of upwards of twenty-four new Orions would give Taiwan the edge it needs to counter the sub-threat. Taiwan could possibly place an order by the end of the century with deliveries by 2001-02.

Besides those nations contemplating new Orion aircraft, there are many more striving to qualify for ex-US Navy surplus P-3A or P-3B airframes through foreign military sales. A number of current P-3 operators acquired their Orions via Navy FMS cases. Others have made inquires and for one reason or another didn't qualify to receive them.

Morocco is one—it once requested baseline P-3As back in the mid to late 1980s. Plans were to have them modified with a P-3C Update II.5 cockpit, IRDS and the Harpoon missile system. The aircraft were to have been used as territorial fisheries protection platforms capable of dealing, although somewhat aggressively, with illegal fishing vessels violating their EEZ. Nothing has been heard from Morocco since August 1990.

The Turkish Navy was offered approximately ten P-3As (eight operational and two spares) in 1991 to offset the acquisition of P-3 Orions to Greece. The Turkish Government later decided to decline the proposed FMS case. The capabilities of the P-3 exceeded their MPA requirements which previously consisted of aging S-2E Trackers. Although no particular follow-on aircraft has been identified, the capabilities of the S-3 Viking are more in line with the Turkish requirements. Turkey has made unofficial inquires to the US Navy as to S-3 availability a number of times, but no decision to part with surplus Vikings at AMARC has ever been made.

Malaysia and the Philippines are currently proposed countries looking at the P-3 for regional coastal patrol and fisheries protection. South Africa has also been mentioned recently as a potential customer of Navy surplus P-3s. Used TACNAVMOD P-3Bs would replace older outdated Douglas Super Dakotas (turbopropped WWII era C-47s) responsible for economic zone protection, sea fisheries, SAR and environmental /oil pollution control, as well as supporting Antarctic region bases and outlining islands.

Since the delivery of UP-3A Orions to Chile, a number of other south American countries are interested in similar ex-Navy P-3s for their forces and are currently making inquires as to aircraft price and availability. They include Argentina, Brazil, Colombia, Uruguay and Venezuela.

Brazil operates aging S-2 trackers and is looking at approximately eight ex-Navy P-3A aircraft. Its neighbors Uruguay and Venezuela are both interested in six used Orions, while Colombia is vying for eight P-3As.

Argentina, of all the Latin American countries interested in Orions, is expected to be the next P-3 operator. Sharing a border with the first South American nation to receive surplus P-3, Argentina has formally requested six TACNAVMOD P-3Bs to replace aging Lockheed L-188 Electras. This recent request is a new inquiry for operational aircraft and not an upgrading of the 1990-91 request for P-3A airframes as systems and engine spares for upgrading its aging L-188 Electras.

Argentina has been interested in P-3s for many years. As mentioned previously, Argentina was to have acquired eight P-3B Orions from Lockheed. These particular aircraft were being traded in by Australia towards the purchase of ten production P-3C Update II Orions. But the Falklands War between Argentina and the United Kingdom caused the Orion deal to fall through with Australia (a commonwealth partner of the UK) refusing to permit the transfer of the aircraft to Argentina. The same aircraft were subsequently sold to Portugal and reworked into P-3P Orions.

Argentina tried again in 1991 when it inquired about several P-3A airframes and parts believed to maintain the operational effectiveness of its Lockheed L-188 Electras utilized

The third world submarine threat ; a KILO class attack submarine like those being transported to Iran.

for maritime reconnaissance and ELINT/SIGLINT operations. The Electras were acquired shortly after the Falklands War and subsequently upgraded in 1993 that included a EXOCET anti-ship missile capability. Argentina's maritime patrol capabilities are split between S-2T Trackers (S-2E upgraded with advanced avionics and Garrett TPE 331-15 Turboprop engines with five bladed Hartzell propellers) for ASW and new Raytheon Beech MPB200T (Super King Airs) equipped with search radar and SAR equipment. These aircraft are tasked with coastal patrol and EEZ protection as well as search and rescue.

Since the delivery of eight UP-3As to Chile, Argentina decided to try once more with all indications pointing to them succeeding. As the saying goes, "third time's a charm!"

One important criteria of any FMS case in Latin America is the concern over organized logistical support. For Argentina and the other potential operators the logistical concerns could be alleviated by the recent acquisition of an Argentinean aircraft depot facility by Lockheed Martin. Area de Material Cordoba, the Argentine aircraft factory and maintenance depot located at Cordoba (100 miles northeast of Buenos Aries) was acquired by Lockheed Martin's Ontario, California, division on March 1, 1995

The new division, Lockheed Aircraft Argentina South America (LAASA) took over the facility under a twenty-five year concession to manage and operate the depot. The privatization of the depot is in conjunction with other signed contracts to refurbish and modify thirty-six A-4M Skyhawks and also provide additional aircraft maintenance services to the Argentine Air Force, encompassing a multitude of aircraft types.

Lockheed Argentina hopes to attract potential customers in the region and offered them an international aeronautical maintenance center providing depot-level inspections, engine refurbishing, avionics upgrades, pilot and maintenance training, spares provisioning and technical documentation. The

Third World countries are operating over 220 submarines, some are new modern-technology vessels that are quiet and very capable. ***via VPI***

A Typhoon Class Submarine, just as capable today. ***via VPI***

new facility could greatly benefit any potential Argentinean P-3 program, as well as those in the region also looking to acquire surplus P-3 aircraft from the US Navy.

THE THIRD WORLD SUB THREAT

With all the focus centered around the ASUW capabilities of the P-3 with the US Navy and the international MPA Community in recent years, little attention has been directed at those developments being made in ASW.

Proliferation of advanced submarines technology to third world maritime countries equates to new challenges for the P-3 Orion to protect the surface fleet battle group forces and their capability to operate forward deployed in various regions of the world. This is a must to support the new US National Security Strategy, as well as being critically important to ensure the United States' ability to protect vital sea lanes of communications (SLOC), exercise sea control and project Naval power abroad under the power projection mission.

The serious threats from third world nations consist of their acquisition of former soviet nuclear submarines or low-cost high-technology diesel submarines equipped with advanced propulsion systems and weapons capabilities—technology that includes Air Independent (diesel) propulsion, and high-density batteries to reduce the need for snorkeling, as well as increasing submerged endurance and weapons such as cruise missiles, advanced torpedoes and anti-ship missiles that all can be launched via conventional torpedo tubes.

ASW is coming full circle, utilizing techniques of the past to combat the submarines of the future. Those capabilities used to counter submarines like this OSCAR II must receive high priority upgrading to combat the next-generation subs of the 21st Century. ***via VPI***

Third world sub tactics can include underwater mine laying, anti-surface and anti-submarine warfare attacks, as well as utilizing the sub as a commando swimmer delivery system. This encompasses maritime terrorism—assaults on merchant shipping, combatant surface ship and logistical support vessels and tenders operating in a littoral environment during contingency and limited objective warfare (CALOW) operations or conflicts. The third world sub is a formidable threat in future littoral operations and potentially more demanding than those operations against the Soviet nuclear boats in the open ocean during the cold war.

The new VP community has already recognized this situation for itself and has in recent years stepped up methods to counter this emerging threat through the establishment of improved ASW training programs. The new training program seeks to keep the existing ASW skills of P-3 crews at peak efficiency in the wake of the cold war and establish the ASW training improvement program. This program stresses methods of detecting both advanced diesels and older nuclear submarines in the shallow water environment. The program considers these submarines' characteristics and operating parameters in the brown water environment. ASWIP provides the means to differentiate contacts from the enormous amount of ambient surface clutter and deals with the sound propagation —bottom scatter noise disruptions that are common to the shallow water environment. The intensity of the ASWIP training program is reminiscent of the ASW training against the first Soviet nuclear submarines during the cold war.

Even the US Navy's force structures have been changed to counter the new third world sub threat. Previously, Navy force structures were exclusively based on the Soviet's ideologies and its cold war doctrines against the Soviet's blue water fleet. Now, changing structures are based on littoral joint operations, focusing on the independent national self-interests of a number of counties around the world. It used to be that you knew who your enemy was. In the post cold war it's not always clear who the enemy is—with possible bad guys everywhere.

Even old enemies surprises the US still. In a recent incident, a Russian Typhoon-Class submarine fired an inter-continental ballistic missile from under Arctic ice as a test. That's why international NATO MPA forces participate each year in ICEX exercises. Under ICEX, MPA forces practice delivery of sonobuoys and ASW weapons where applicable through tiny fissures and cracks in the ice pack in order to track and attack submarines operating under the Arctic ice sheet.

As ASW comes full circle and new emphasis is placed on ASW in future operations of the P-3 Orion, a series of development programs have been established to exploit existing

Lidar Laser equipped P-3C being tested for future Littoral "shallow-water" ASW applications. ***US Navy / NAWC Warminster - Patuxent River***

acoustic technologies and produce new ones to combat the undersea threat.

They include improvements to existing passive sonar systems. Although increasing sensitivity is the principle objective, it may not be enough to re-capture those ASW capabilities lost due to the advanced technology of newer submarines—which diminish the amount of acoustic energy generated.

The future may lay in the re-emergence in very-low-frequency, long range active sonar with longer active detection ranges. This equates to higher search parameters and greater detection probabilities. Low Frequency Active Acoustics provides reliable submarine detection at ranges far exceeding those of current passive acoustics with no dependence on radiated noise.

One active acoustic program, to improve the detection of advanced submarines is the Extended (Explosive) Echo Ranging, or EER. EER has shades of the old Jezebel/Julie active acoustic methodology of the P-3A in 1963.

In the next twenty years, future acoustic improvement programs could include nullification of ambient noise and the adaptability of acoustics to specific local environments requiring previous survey and characteristics of that local littoral environment recorded.

New advanced non-acoustic detection sensors are also a priority of future ASW operations and include "Lasers" for detection and ranging. Development of lasers as a ASW detection sensor has been on-going for many years with the P-3 being an integral part of that developing process. Back in the mid 1980s, depth penetrating lasers were considered as a possible submarine detector. The ASW Laser research program was established to test the parameters of lasers in the maritime environment to detect undersea objects. NAWC-AD Warminster performed development test flights with test range Orions specially modified for laser projects.

The prototype light detection and ranging laser (LIDAR) combines fiber-optics and laser technology to produce the

NAWC Pax River (NATC) P-3C testing new MK50 Torpedo. ***V. Pugh via US Navy / NAWC Patuxent River***

NAWC Pax River (NATC) P-3C test firing advanced, integrated Maverick missile system. ***V. Pugh via US Navy / NAWC Patuxent River***

depth penetrating laser energy to detect objects in shallow water or the deep water environment.

As late as December 1995, the Naval Research Laboratory took up the challenge by flight testing an Australian subsurface laser ranger on board one of their NP-3D Orions to establish operating parameters for Navy based ASW lasers in the future.

Other laser projects have also been developed and tested aboard NAWC Warminster P-3 aircraft. These undersea laser applications dealt with using lasers as the means for satellites to communicate with submerged submarines.

US Navy proposed CEC test and development platform for NRL. ***Lockheed***

CEC test aircraft; note weapons bay mounted antenna radome. ***Lockheed***

The Tactical Airborne Laser Communications Program was made up of two test programs, one known as project "Y" Blue and Green. This project utilized lasers of different wave lengths to demonstrate one-way (downlink) communications from a P-3 to a submerged submarine. Project "AOR" or Advanced Optical Receiver tested the reverse demonstrating a P-3 Orion aircraft could receive a laser (up-link) from a submerged submarine. Both these initial test proved successfully in their defining maritime laser parameters. These projects also gathered various information on cloud thickness, different sea conditions, aircraft altitudes and water depths parameters that might effect the operational effectiveness of the LIDAR lasers.

WEAPONS

Future ASW operations also require new and enhanced weapons for the P-3 to counter advanced (quieter) submarines. The Mark 46 Improvement Program includes a SLEP of the existing anti-submarine torpedo to improve its shallow-water capabilities as an interim before the introduction of the new Mark 50. The MK50 torpedo is capable of shallow water ASW attack with evasive targeting mode to combat submarine counter measures. There is even a Harpoon anti-ship missile upgrade encompassing a new GPS navigation and targeting re-attack mode.

CEC

In the future the P-3 is expected to take on more mission responsibilities. In conjunction with the overall protection of the battle group, the P-3 will be inducted into the Cooperative Engagement Capabilities program. Under CEC, the P-3 is to be equipped with interactive electronics, linking the aircraft with on-board ship sensors to provide the battle group with a Ballistic Missile Defense capability.

The P-3 will provide over-the-horizon range coverage that increases the battle group OTH detection range, thereby enhancing its anti-air warfare capability. Although originally to have been incorporated into the carrier-based E-2C Hawkeye, the CEC program now includes equipping the P-3, other tactical airborne assets and land based stations to produce an AAW umbrella over the battle group to guard against ballistic missiles and anti-ship missile attacks.

The P-3 has also had a hand in development of the CEC program. NAWC Pax River, tasked with providing proof-of-

Courtesy of *Baldur Sveinsson*

concept, flight testing of the prototype CEC system, utilized a borrowed P-3 AEW&C from US Customs Service to test what had been a ship-borne system.

Due to the need for a dedicated test aircraft and US Customs' busy schedule, NRL was selected to carry on development test of the CEC program. NRL has modified one of its test range NP-3D Orions with a twenty-four foot rotodome much like the Customs domes. NRL will also operate the aircraft for E-2C radar development and other ballistic missile theater defense capabilities tests.

Appendix A
Major P-3 Variant Descriptions

P-3A "ALPHA" - The first ASW configured model of the P-3 Orion, based on the L-188 Electra. (originally designated P3V-1 until September 1962) The P-3A model would later go on to spawn a number of mission dedicated variants.

CP-3A ORION - A proposed cargo / passenger transport version of the P-3A based on a requirement of the US Navy. A contract was let to Lockheed for development of the aircraft, but was later canceled and the Orion never furthered beyond the drawing board.

EP-3A ORION - Originally designated for the testing of a new airborne electronic intelligence gathering system, this model ended up becoming a catch-all designation for a number of different aircraft involved in special mission and electronic systems test projects. Each was modified uniquely according to that task.

EP-3E ARIES - A group of P-3 Alphas modified with the ARIES electronic intelligence gathering systems and utilized in the electronic reconnaissance role.

NP-3A ORION - A designation given to an aircraft that has been so drastically modified from its original mission configuration, that it is beyond any practical economic limits to restore it to its former use. The best representation of this is the NP-3A once own and operated by NASA.

RP-3A ORION - A designation given to Orions specially modified for oceanographic research.

TP-3A ORION - A designation given to a number of Orions equipped as pilot trainers.

UP-3A ORION - A designation given to a number of specially modified Orions operated in the utility transport role as an low cost alternative to the CP-3A. This designation is shared by a group of Orions individually modified for various airborne systems and weapons tests.

VP-3A ORIONS - A designation given to a group of specially modified Orions utilized for VIP transport.

WP-3A ORIONS - A designation given to a group of Orions modified for US Navy weather reconnaissance or "hurricane hunting."

YP-3A ORION - The P3V-1 / P-3A prototype aircraft which was originally developed from a L-188 Electra. The aircraft was later transferred to NASA as an NP-3A

P-3B "BRAVO" - The second ASW P-3 model to be developed, designed as a interim aircraft between the analog technology of the Alpha and the digital sophistication of the P-3 Charlie Orion. Again, the Bravo model became the baseline for a number of variants.

EP-3B ORION - Known as "BAT RACKS" the two aircraft were the initial version of the P-3 based electronic intelligence gathering aircraft. Although built from two Alphas the aircraft received a "B" designation - they were both later brought up to EP-3E ARIES standards. This designation was also shared with one lone Bravo aircraft specially modified as a EW simulator / evaluator flying laboratory.

NP-3B ORION - A semi-official designation for a uniquely modified P-3B utilized for special operations taskings.

UP-3B ORIONS - The utility transport version of the P-3B replacing retiring UP-3A.

P-3C "CHARLIE" - The third ASW/MPA model of the Orion to be produced with digital electronic sophistication. This model was improved upon several times through a series of "Updates" - each seeking to keep pace with advances Being made in Soviet submarine technology.

P-3F ORIONS - A designation given to a group of Orions produced for IRAN that were based on P-3C airframes with P-3A/B avionics.

P-3G ORION - A temporary designation used prior to the official designation of the proposed LRAACA / P-7A follow-on ASW platform to the P-3C.

P-3H ORION - A short-lived designation for the proposed ORION II follow-on ASW platform to the P-7A.

P-3K ORIONS - The designation given to a number of P-3 Bravos acquired by New Zealand and later improved to the point of the appointment of a new designation - to reflect the uniqueness of the aircraft's configuration.

P-3N ORIONS - A designation given to two specially modified Norwegian P-3B utilized in coast guard duties and utility transport taskings.

P-3P ORIONS - A designation given to a group of Orions based on ex-Australian P-3B airframes acquired by Portugal and re-equipped with P-3C Update II avionics.

P-3T ORIONS - A designation given to two modified P-3A acquired by Thailand. (UP-3T) - A special one aircraft utility version of the THAI "P-3T."

AP-3C ORION - A proposed designation for a number of Australian P-3C Update II.5 that have been improved via a domestic Australian upgrade program.

EP-3 ORIONS - a designation given to a variant aircraft of the Kawasaki HI / JMSDF P-3C Orion , designed as an electronic intelligence gathering platform.

EP-3E ARIES II - A group of P-3C (NUD) airframes specially equipped with the improved ARIES II electronic intelligence gathering reconnaissance systems.

EP-3J ORIONS - A designation given to two US Navy P-3B re-configured into C³CM counter-measures training aircraft.

NP-3D ORIONS - A designation recently adopted for a group of Orions with individual mission taskings and capabilities.

RP-3D ORIONS - The designation originally given to a production P-3C airframe specially converted on the production line for airborne magnetic surveys. A group of specially modified P-3 Bravos also share this designation having been re-configured as replacements for the RP-3A - all these Orions were recently re-designated "NP-3D."

UP-3C ORION - A designation given to a Kawasaki production P-3C developed as a flying test and evaluation aircraft for new airborne electronic sensors and systems for the JMSDF Orions.

UP-3D ORIONS - A designation given to a couple of Kawasaki production P-3C specially developed as C³CM fleet training aircraft with a mission somewhat like the American EP-3J.

UP-3E ORIONS - The forth Kawasaki / JMSDF P-3C Orion variant to be based on existing (ASW configured) P-3C airframes - converted as special mission reconnaissance platforms.

WP-3A ORIONS - A designation given to two production P-3C airframes converted on the production line for weather reconnaissance or "hurricane hunting" with

TP-3C ORION - A proposed US Navy designation for some heavy weight P-3B or P-3C NUD Orions to be modified with P-3C Update III+ cockpits and converted for use as pilot trainers - replacing older TP-3A Orions.

TAP-3 ORIONS - A designation given to a number of specially modified ex-US Navy P-3 Bravos acquired by Australia for RAAF Orion pilot training / logistics transport aircraft.

YP-3C ORION - A designation given to one P-3B airframe converted on the production line into the P-3C prototype aircraft. The Orion has since been re-converted into a RP-3D - having recently been re-designated a NP-3D.

CP-140 AURORA - A Canadian designation given to a number of production P-3C airframes uniquely equipped with modified S-3A Viking sensors, systems and avionics.

CP-140A ARCTURUS - A Canadian designation given to a variant group of Aurora based on P-3C Update III airframes - devoid of sensor systems.

P-3 AEROSTARS- An unoffical designation for a group of surplus P-3A Orion airframes converted as "air tankers" for the purpose of airborne forest fire fighting.

P-3 AEW&C- A designation given to a group of P-3B airframes re-configured as early warning and control aircraft employed in the war on drugs - by the US Customs Service.

Appendix B
Aircraft Specifications and 3-way Diagram

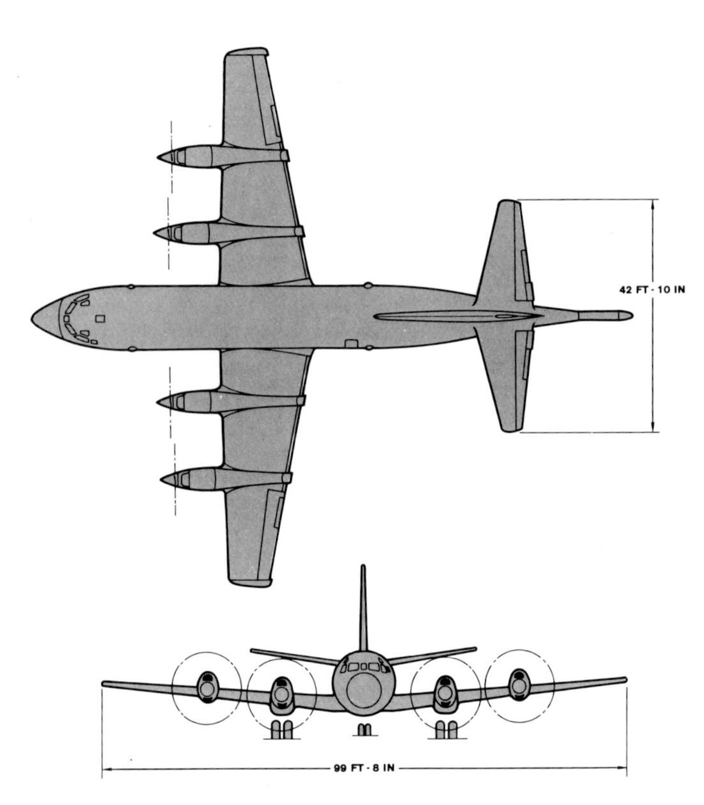

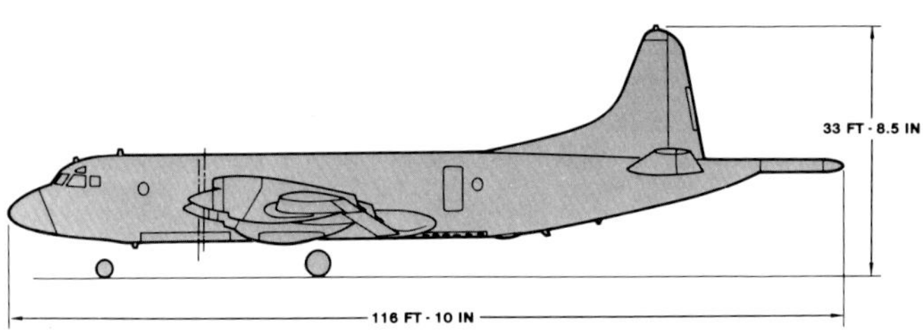

P-3A Orion 3-way chart and specifications ***Lockheed***

Crew: 10-12

Power Plant: Allison T56-A-10W Turboprop Engines

Performance: Max Speed - 380 knts (703 km/h) or 437 mph @ 15,000 ft (4570m)

Ceiling - 28,300 ft (8625m)
Max Range - 4700 n. mi. (7660m)

Dimensions: Span - 99 ft, 8 in (30.38m)
Length - 116 ft, 10 in (35.61m)
Height - 33 ft, 8.5 in (10.27m)
Wing Area - 1300 sq ft (120.77 m^2)

Armament: Max Weight - 7250 lbs (3289 kg) weapons bay
- 16,000 lbs (7257 kg) 10 underwing pylon stations

P-3C UPDATE III & CP-140 CHARACTERISTICS

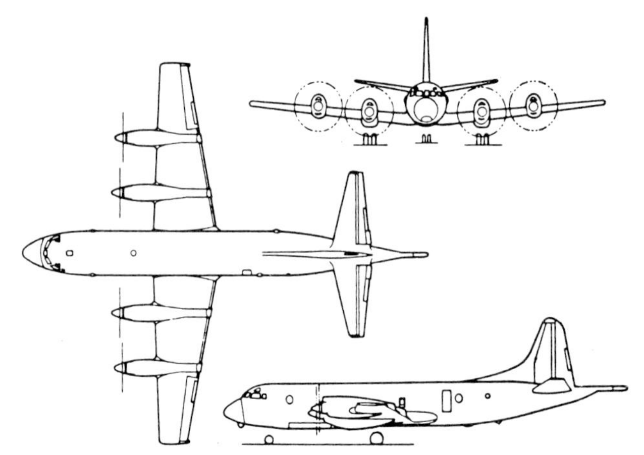

The P-3C and the CP-140 share the same airframe and engines, but their mission requirements differ. When such differences affect the aircraft characteristics, those elements unique to the CP-140 are noted in parentheses.

GENERAL INFORMATION

Fuselage length	116 feet, 10 inches
Wing span	99 feet, 8 inches
Wing area	1,300 square feet
Tail height	33 feet, 9 inches
Maximum gross weight	142,000 pounds
Maximum landing weight	103,880 pounds
Design zero fuel weight	77,200 pounds
Usable fuel capacity	62,560 pounds

PERFORMANCE

Takeoff run at sea level	4,240 feet (3,400 feet)
Maximum landing distance	2,900 feet (2,100 feet)
Max. speed at 15,000 feet alt.	380 KTAS (405 KTAS)
LR cruise speed at 25,000 feet	350 KTAS
Loiter speed at 1,050 feet	203 KTAS
Maximum endurance	16 hours (17 hours)
Service ceiling	34,400 feet
Time on station at 1,600 n.mi.	5 hours
Ferry range	4,830 n.mi.

ENGINES

Four Allison T56-A-14 turboprop engines rated at 4,910 equivalent shaft horsepower. Hamilton Standard 56H60-77 propellers, diameter 13 feet, 6 inches.

CREW

Normal complement	10

P-3C Orion and CP-140 Aurora 3-way chart and specifications ***Lockheed***

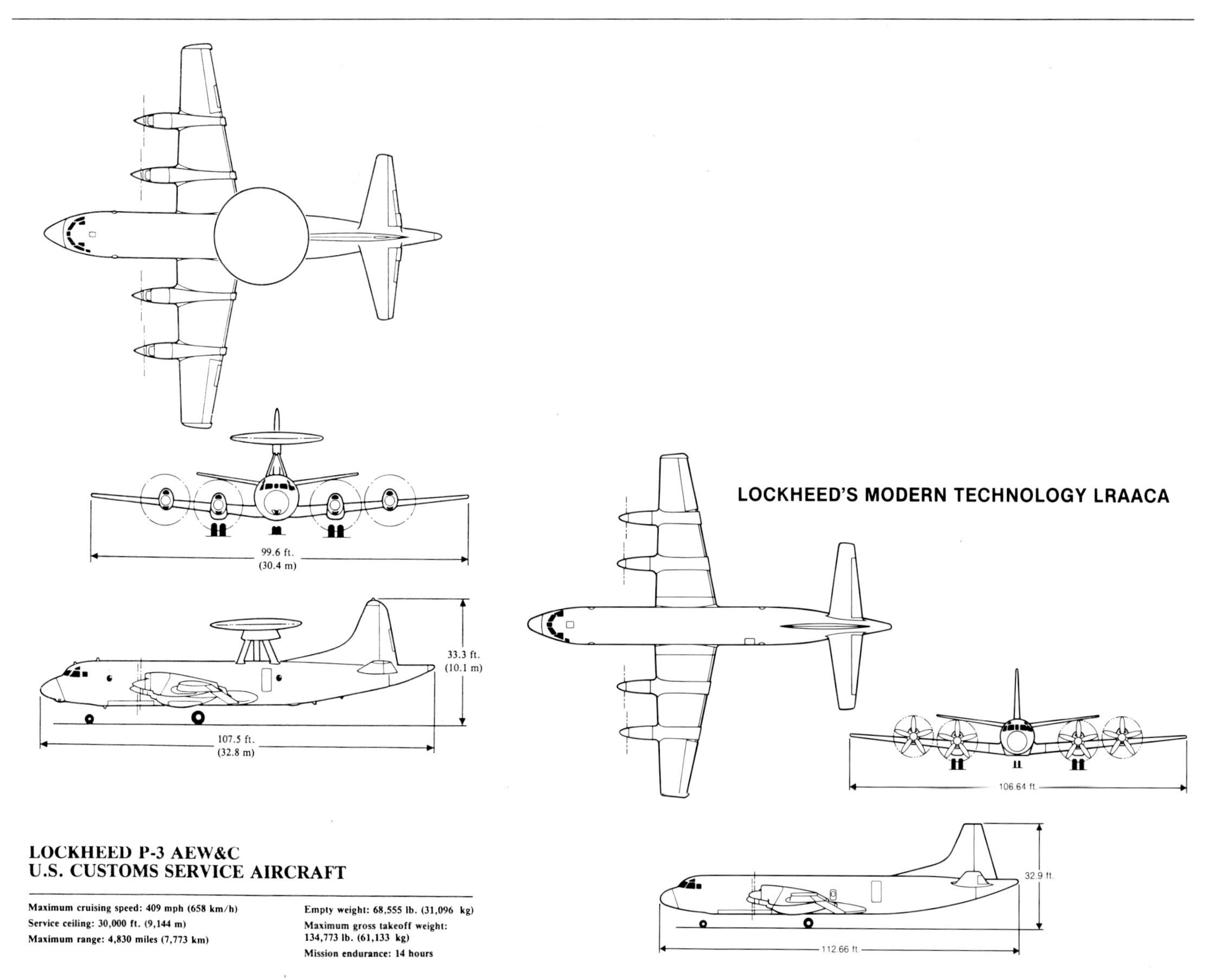

P-3 AEW&C 3-way chart and specifications ***Lockheed*** **P-7A LRAACA 3-way diagram** ***Lockheed***

Appendix C
Major Variant Diagrams and Interior Layouts

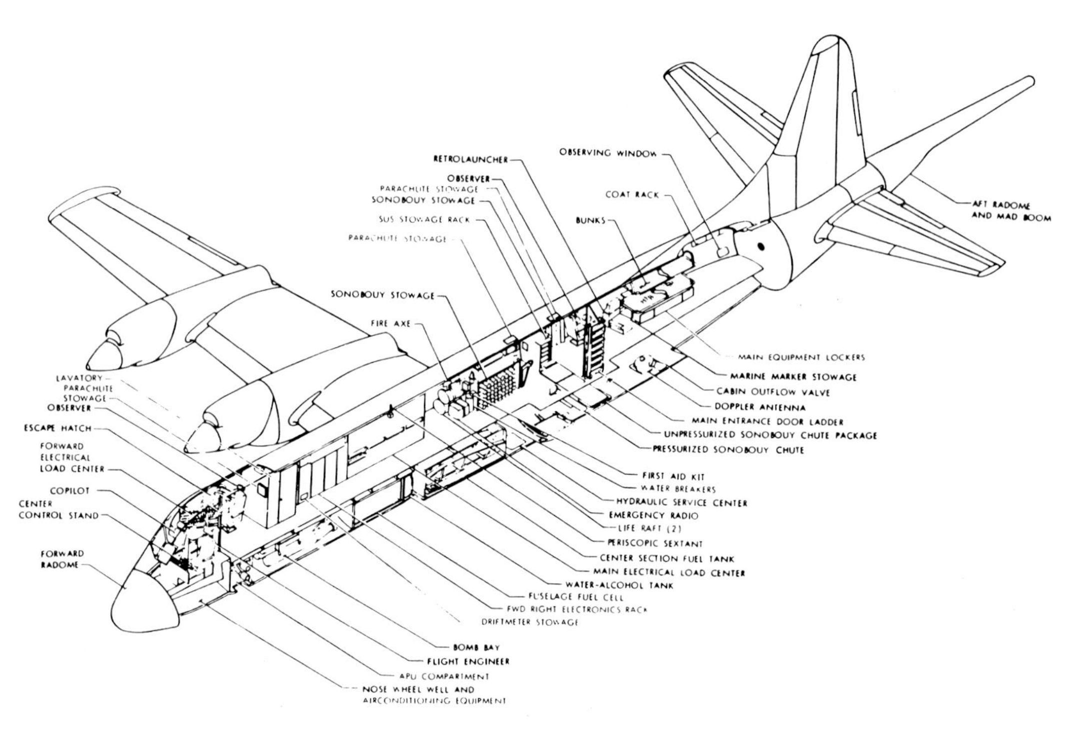

P-3A/B Orion cut-away (right side) ***Lockheed***

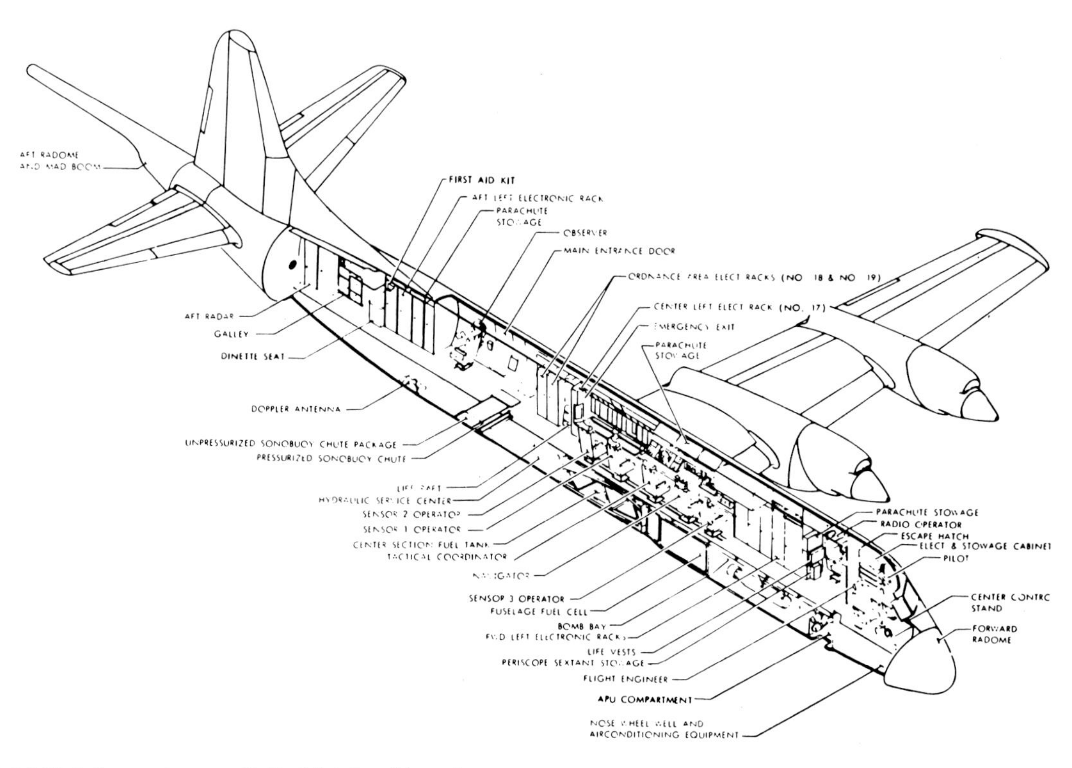

P-3A/B Orion cut-away (left side) ***Lockheed***

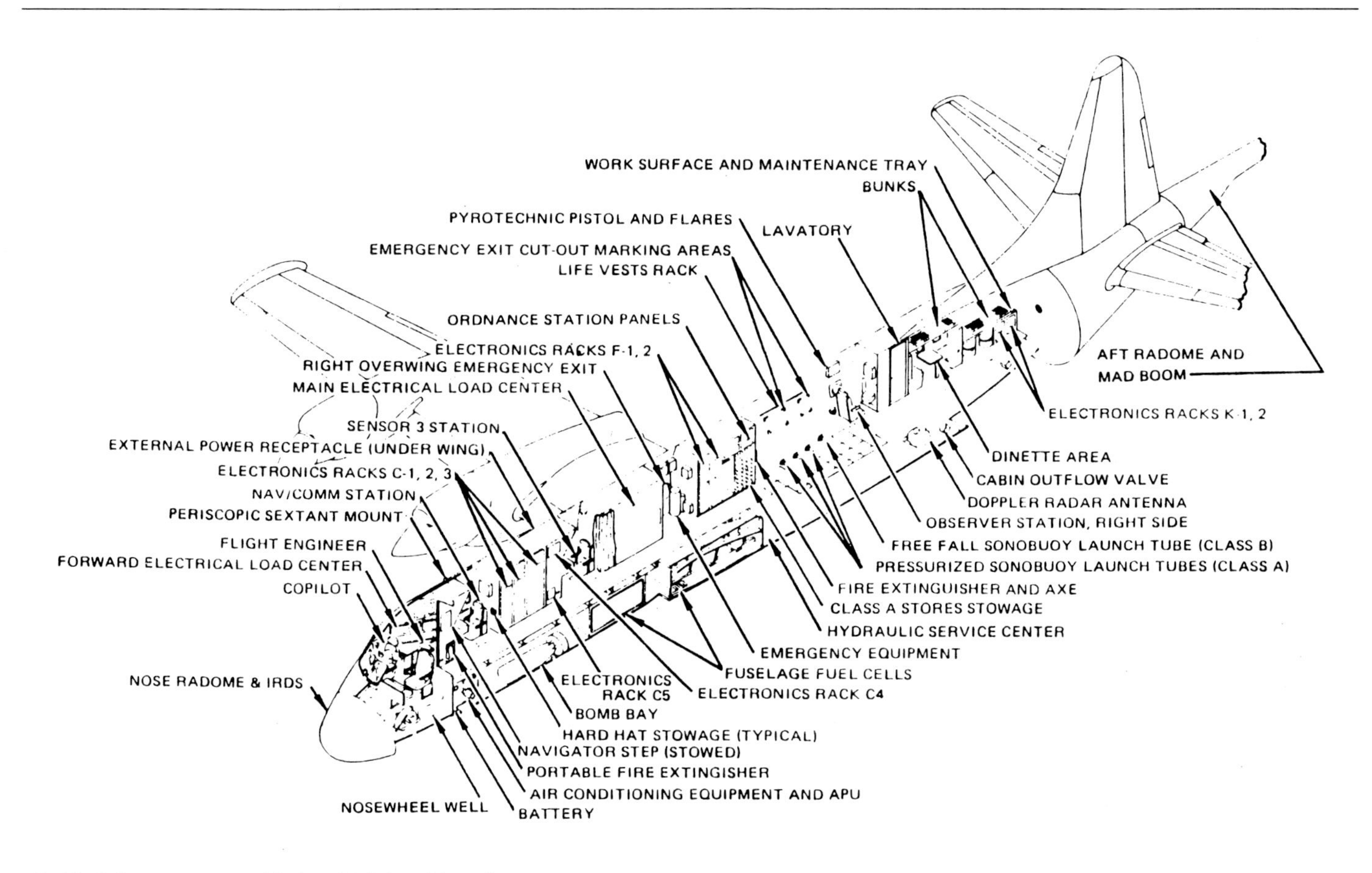

P-3C Orion cut-away (right side) ***Lockheed***

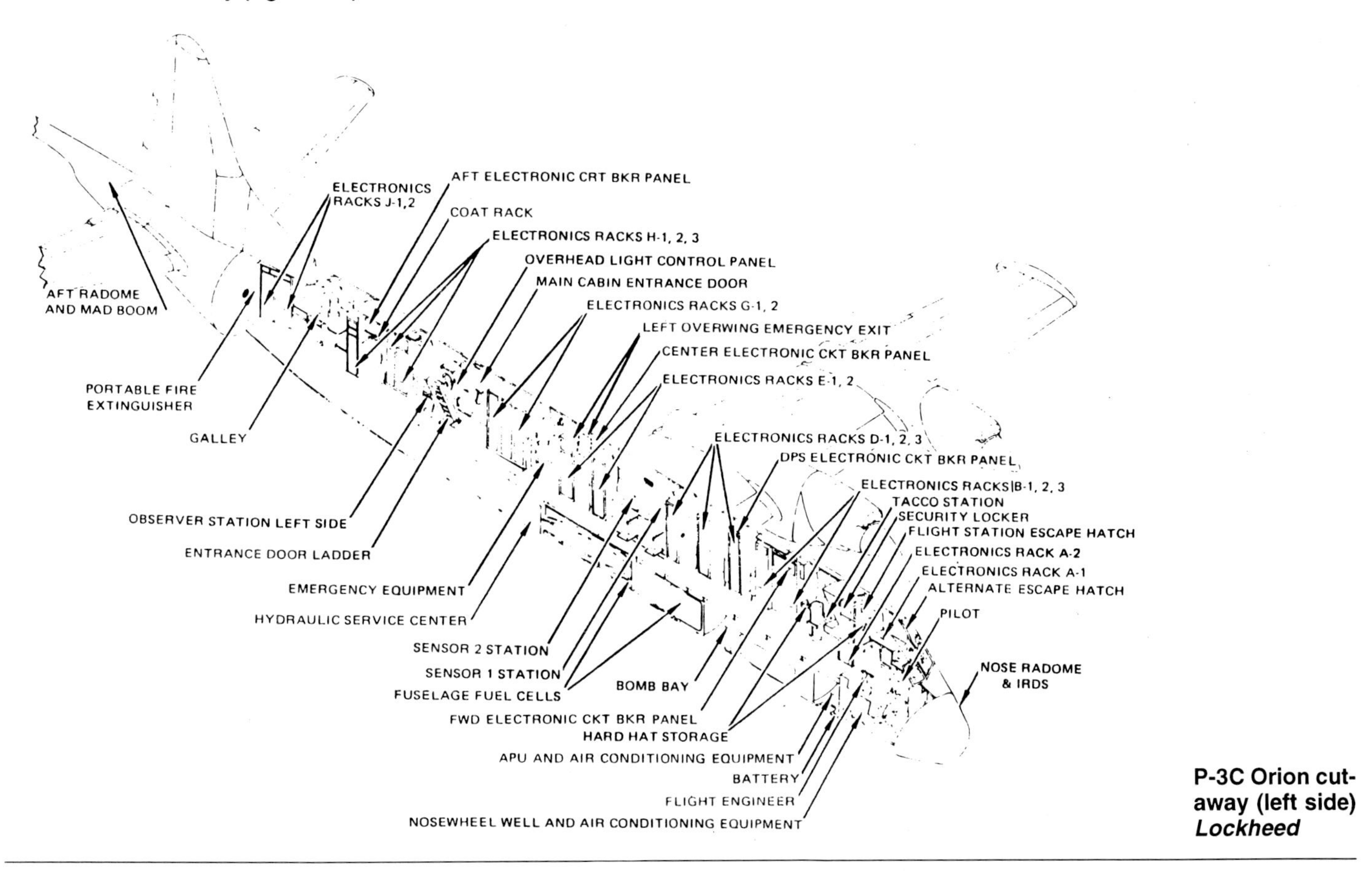

P-3C Orion cut-away (left side) ***Lockheed***

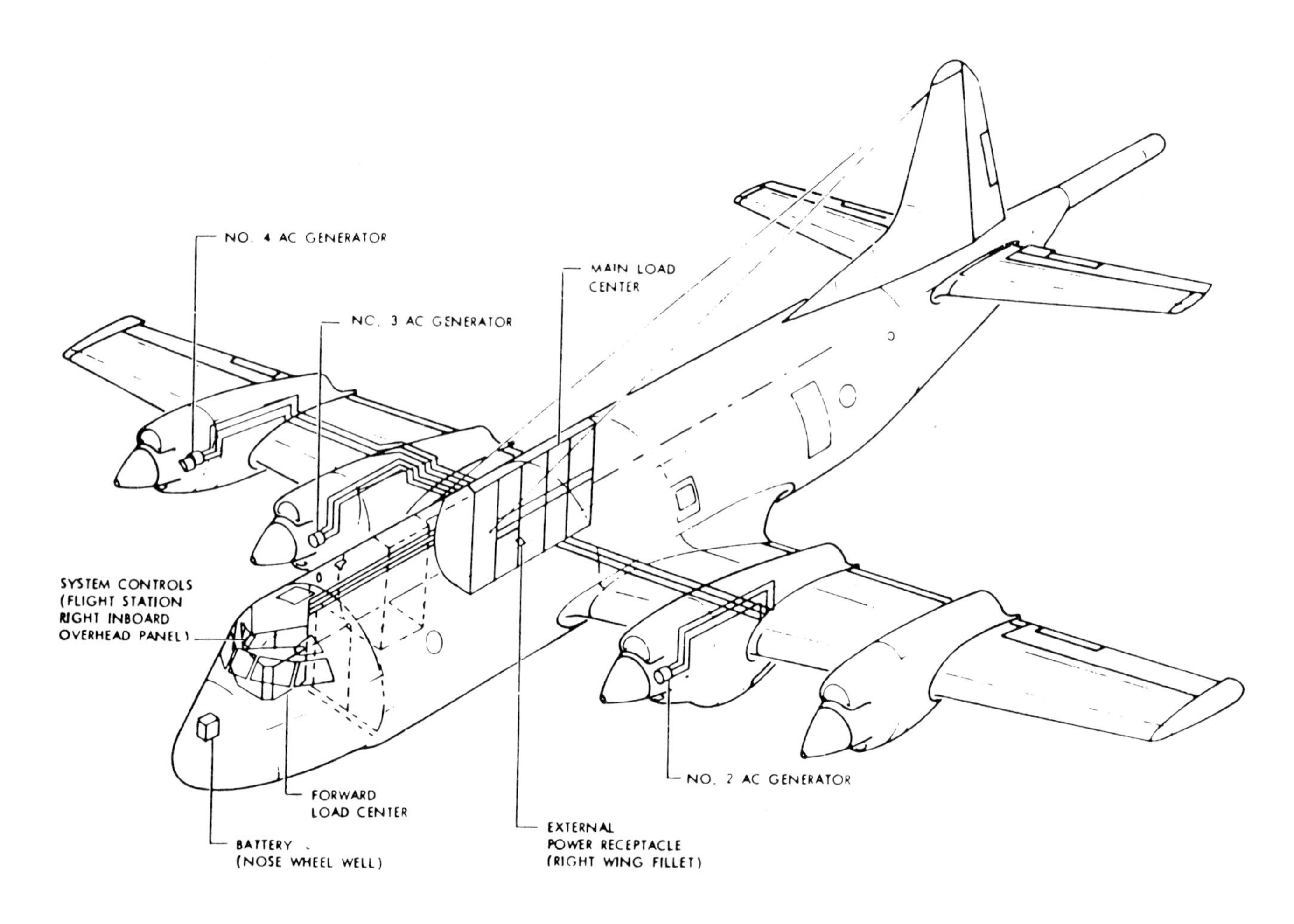

P-3 Orion electrical power system layout *Lockheed*

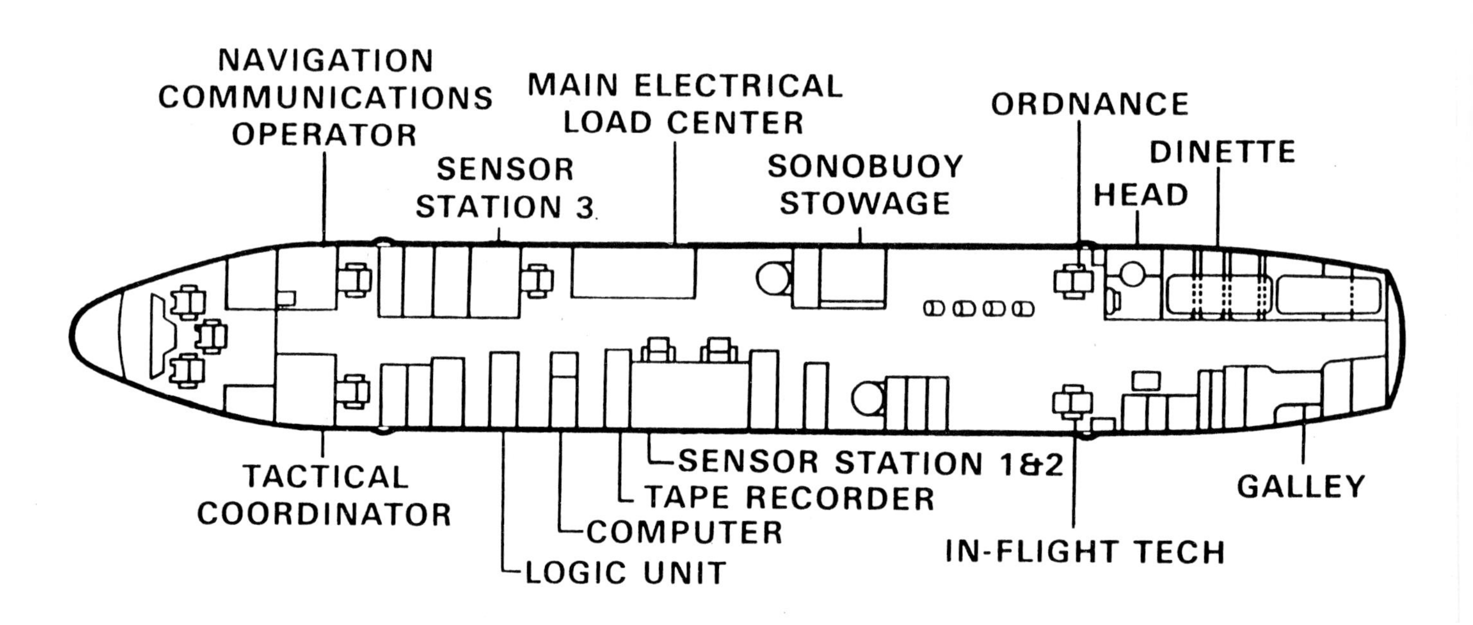

P-3C Orion interior arrangement *Lockheed*

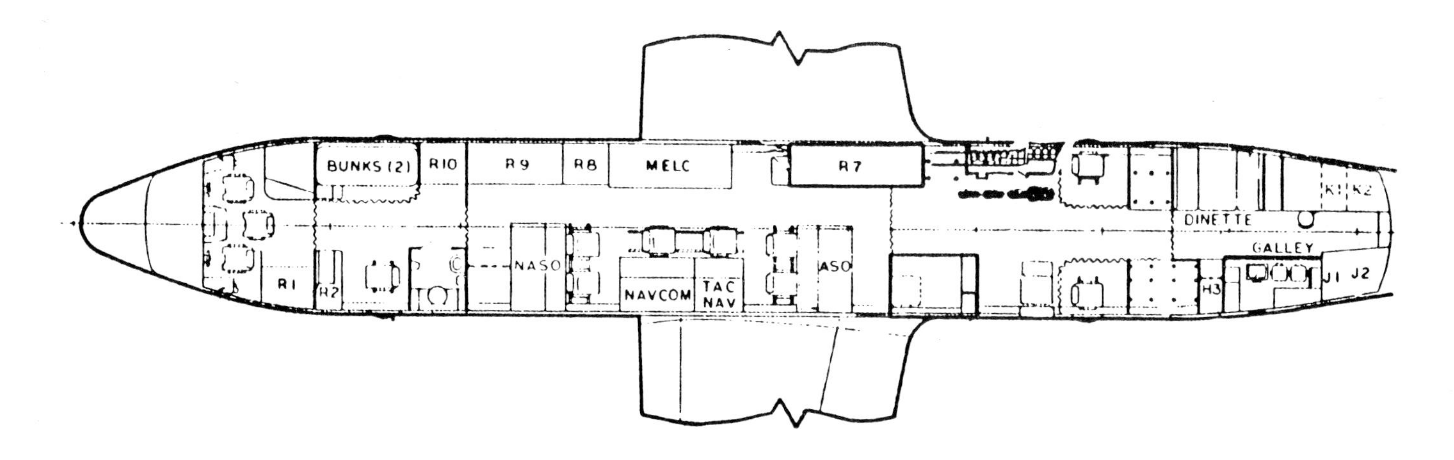

CP-140 Aurora interior arrangement *Lockheed*

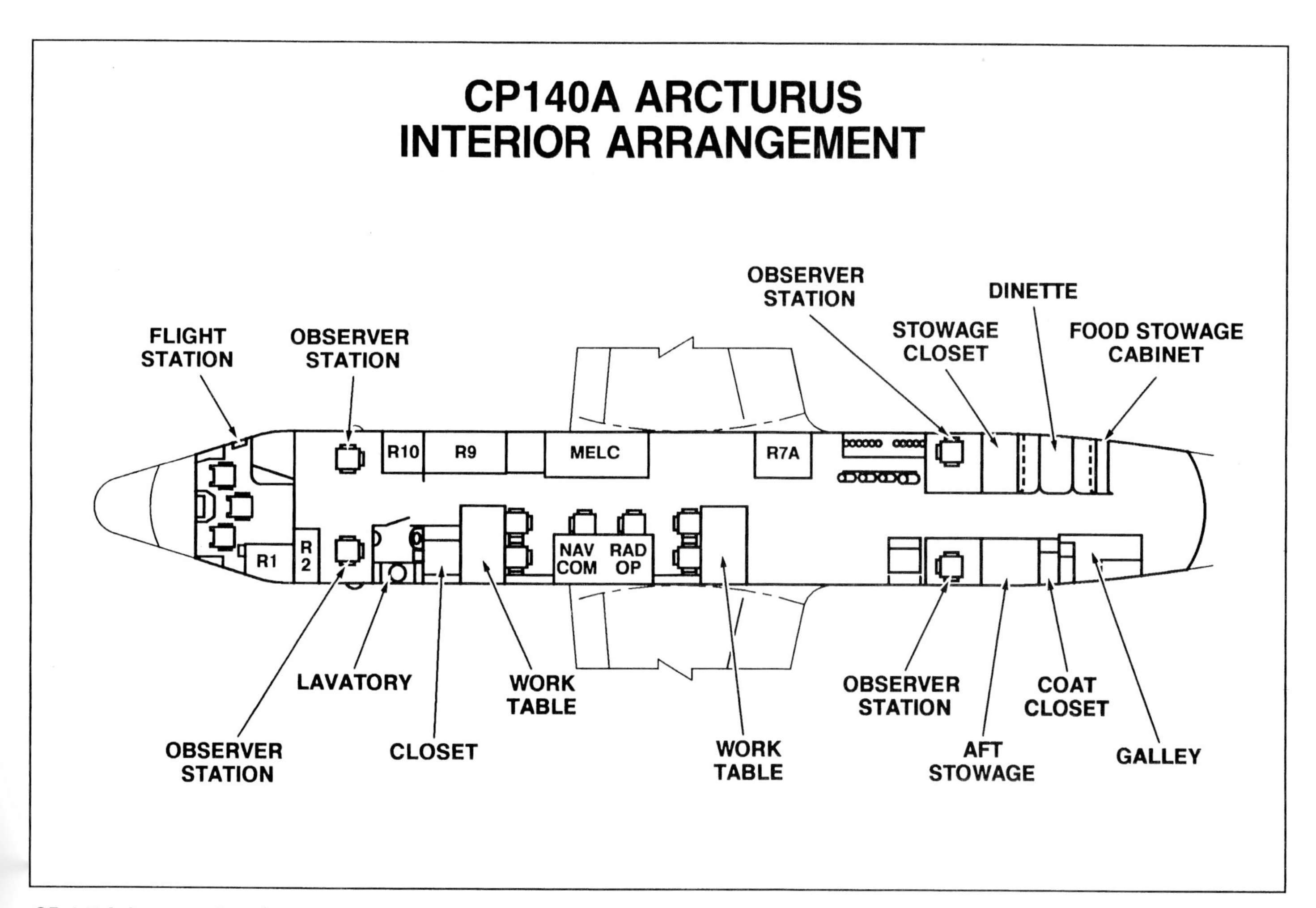

CP-140A Arcturus interior arrangement *IMP*

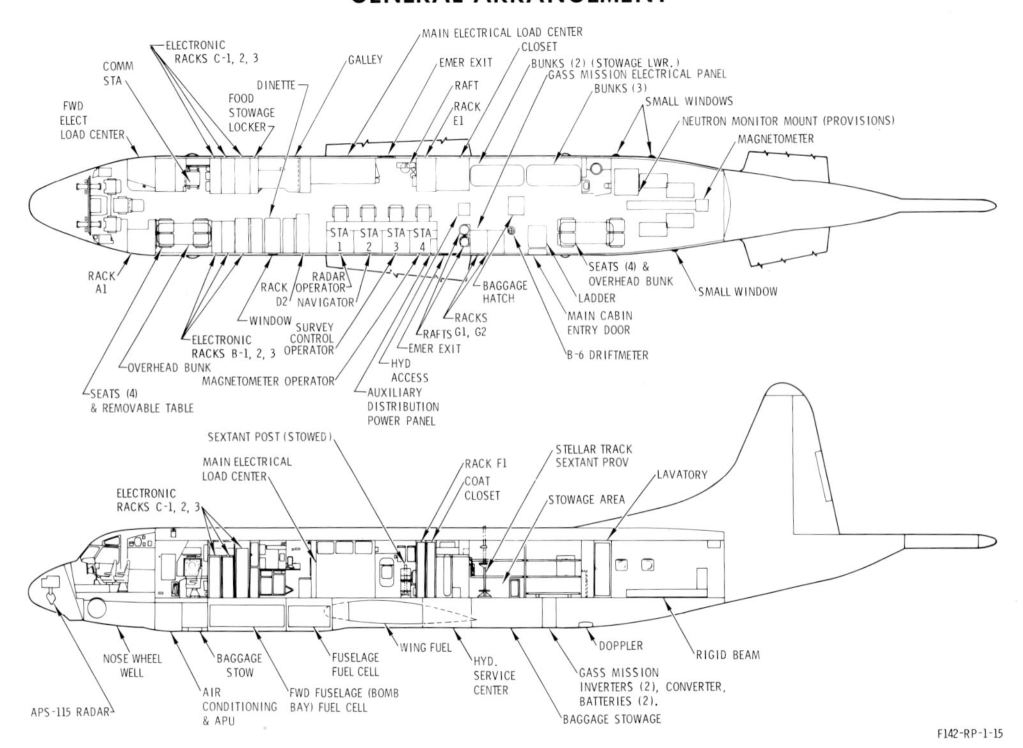

RP-3D Project "MAGNET" Orion interior arrangement. (Lockheed)

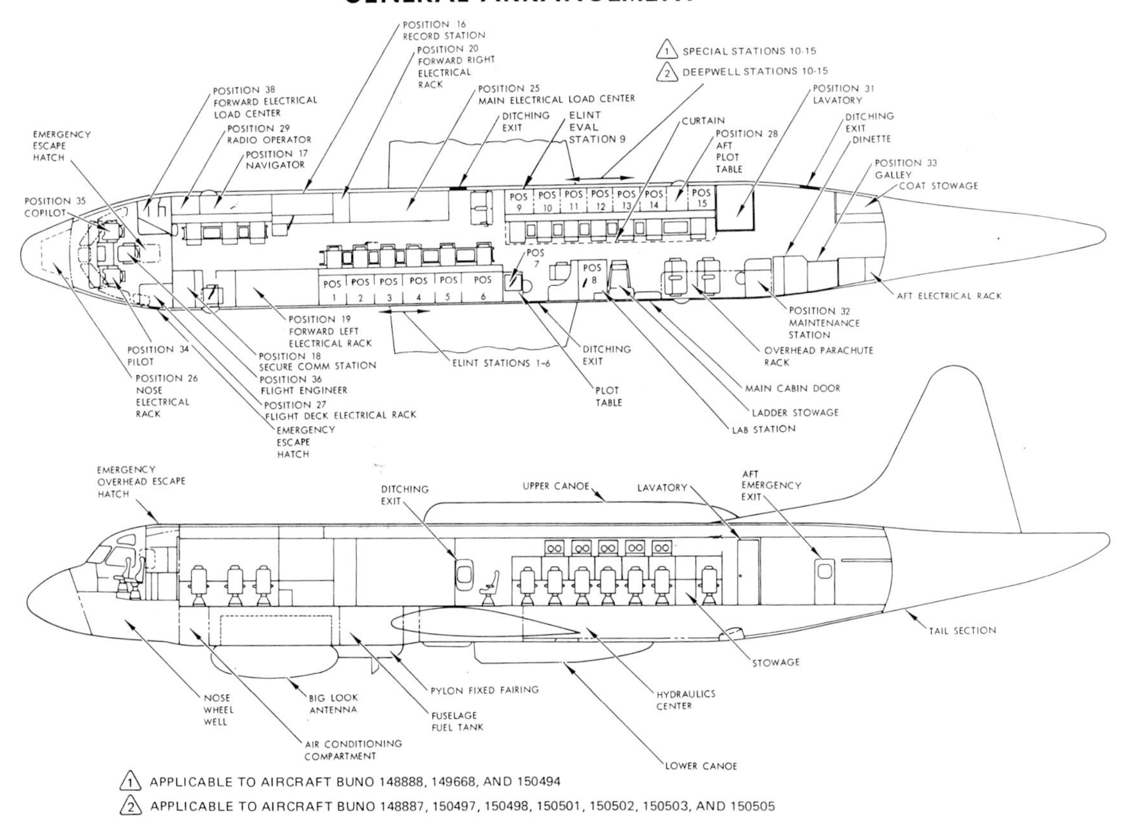

EP-3E ARIES (Deepwell) 2-plane cutaway/interior arrangement. (Lockheed)

NOTES:

RACKS - A - K ELECTRONIC EQUIPMENT RACKS
DS - DITCHING STATIONS
FS - FLIGHT STATION (AIRCRAFT FRAME NUMBERS)
FELC - FORWARD ELECTRICAL LOAD CENTER
MELC - MAIN ELECTRICAL LOAD CENTER

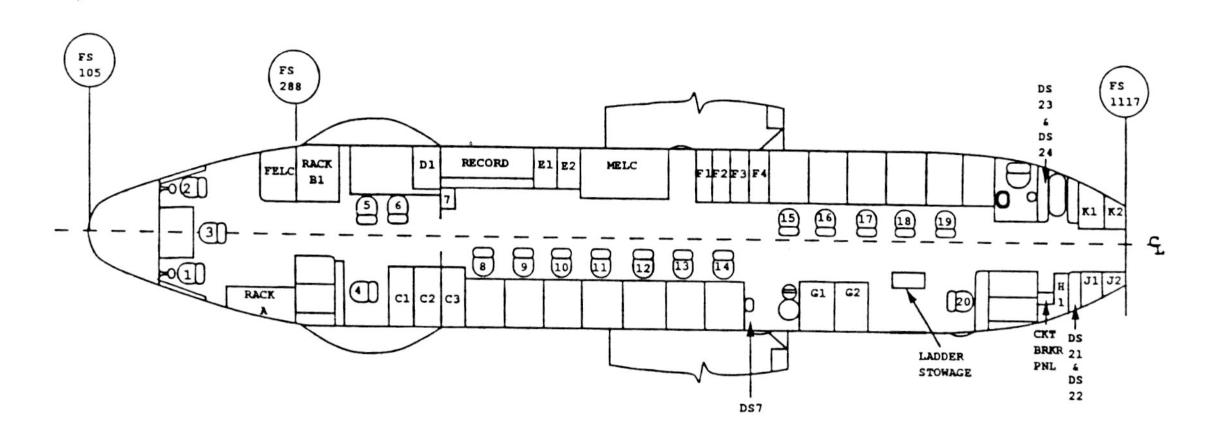

CREW STATION 1 - PILOT
CREW STATION 2 - CO-PILOT
CREW STATION 3 - FLIGHT ENGINEER
CREW STATION 4 - SECURE COMM OPER
CREW STATION 5 - COMMUNICATOR
CREW STATION 6 - NAVIGATOR
CREW STATION 7 - RECORD OPER/FLT TECH
CREW STATION 8 - MANUAL ESM OPER
CREW STATION 9 - BRIGAND OPER
CREW STATION 10 - LAB OPER
CREW STATION 11 - RADAR/ESM OPER
CREW STATION 12 - ESM SUPERVISOR
CREW STATION 19 - SPECIAL SYSTEMS OPER
CREW STATION 20 - S & T OPER
CREW STATIONS 21-24 - DITCHING STATIONS
CREW STATION 13 - EW COMBAT COORDINATOR
CREW STATION 14 - SPECIAL SYSTEM EVAL
CREW STATION 15 - SPECIAL SYSTEMS OPER
CREW STATION 16 - SPECIAL SYSTEMS OPER
CREW STATION 17 - SPECIAL SYSTEMS OPER
CREW STATION 18 - SPECIAL SYSTEMS OPER

EP-3E Aries II interior arrangement. (Lockheed)

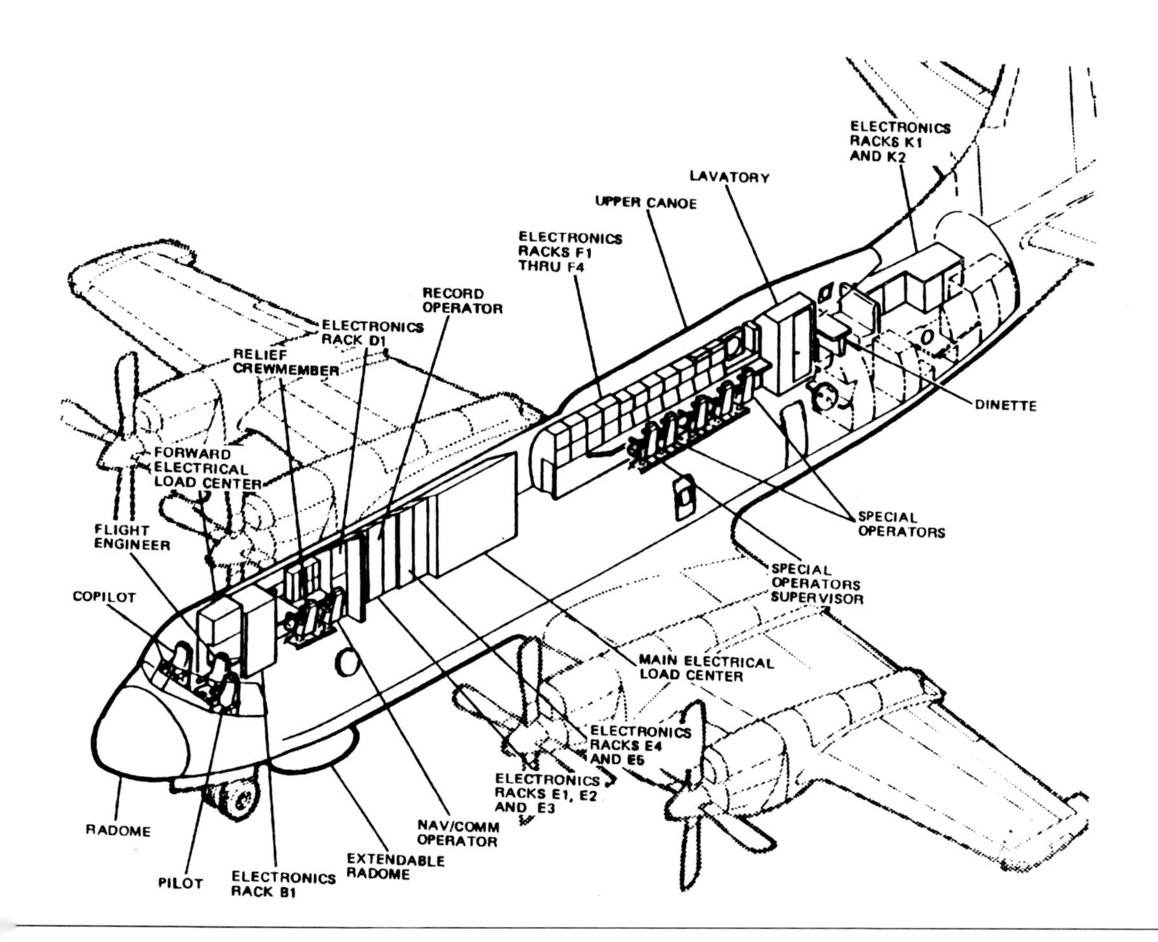

EP-3E electronics cutaway (right side). (Lockheed)

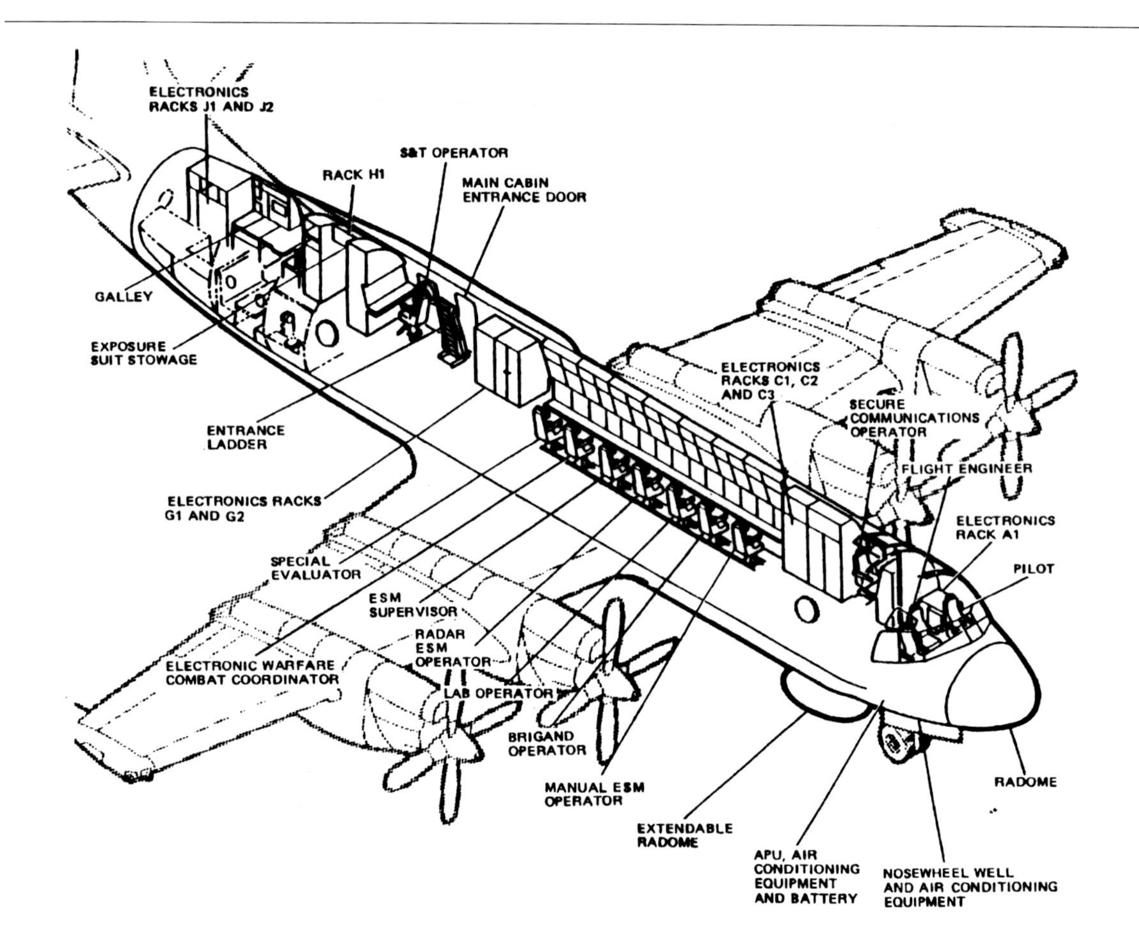

EP-3E electronics cutaway (left side). (Lockheed)

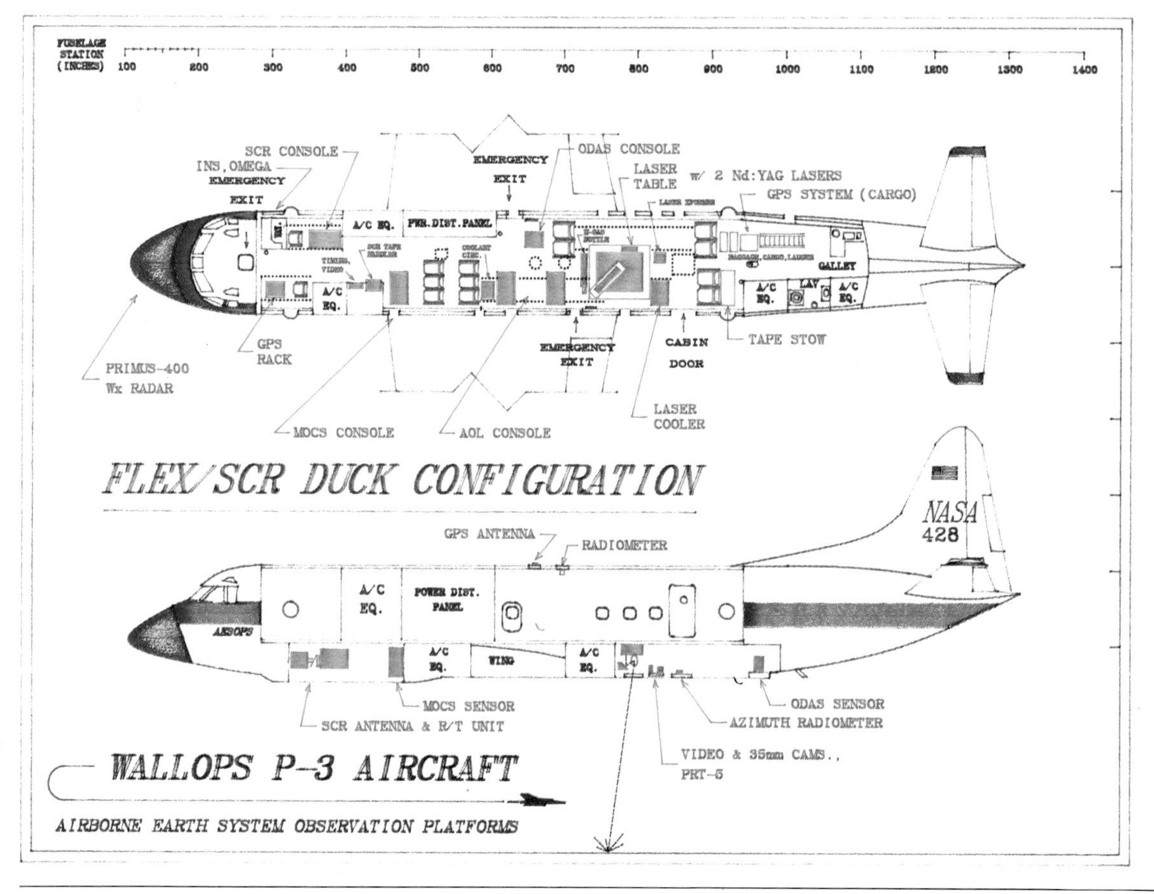

NASA NP-3A 2-plane interior arrangement-side view (fully equipped). (NASA)

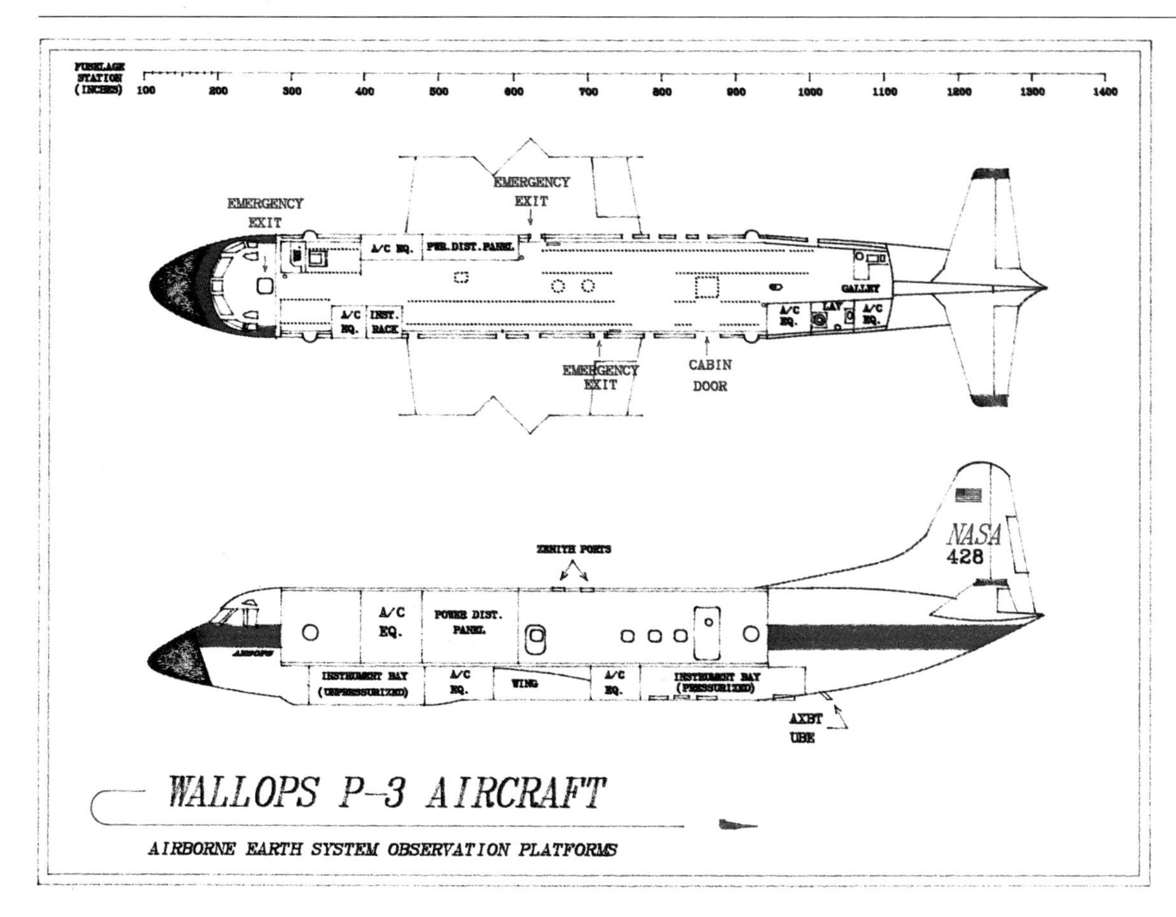

NASA NP-3A 2-plane interior arrangement-side view (fully equipped). (NASA)

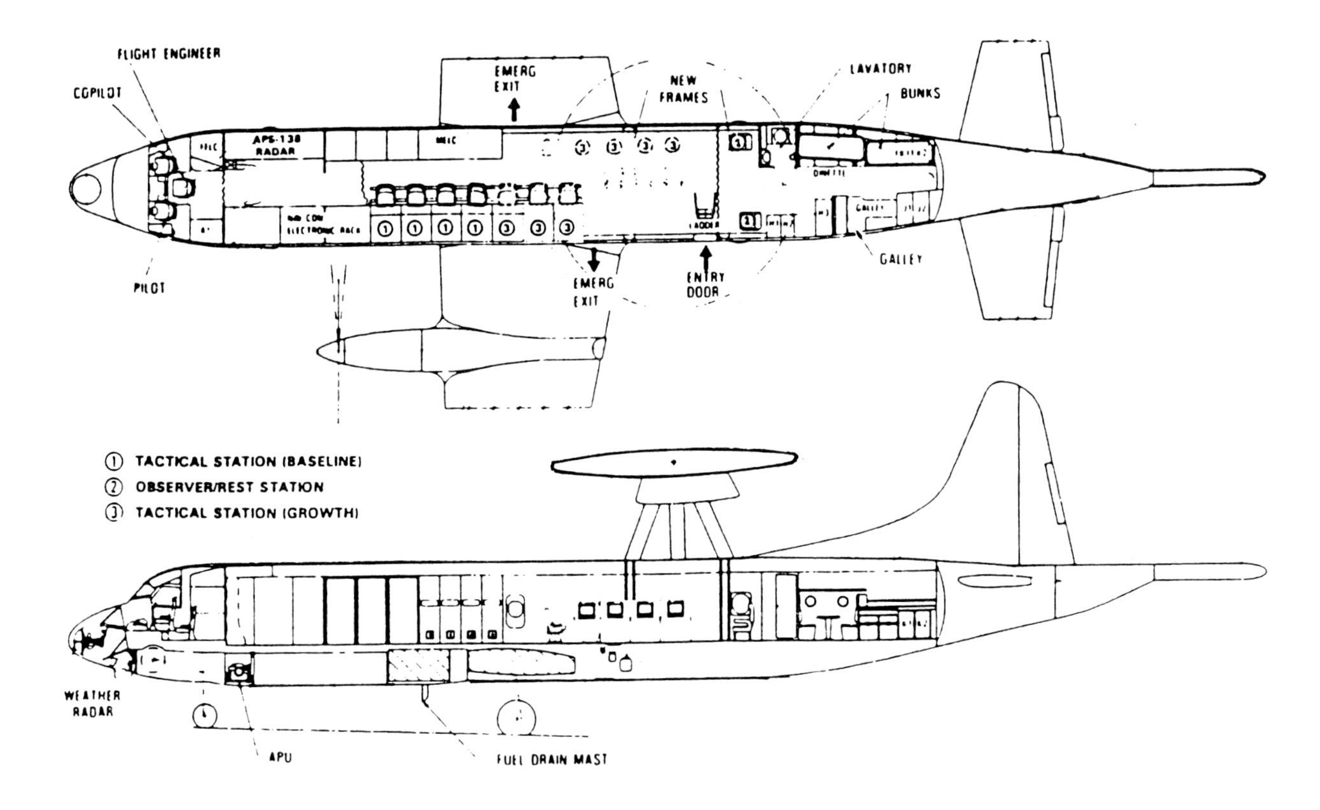

P-3 AEW&C 2-plane interior arrangement-side view (fully equipped). (Lockheed)

Interior Arrangement

NOAA WP-3D

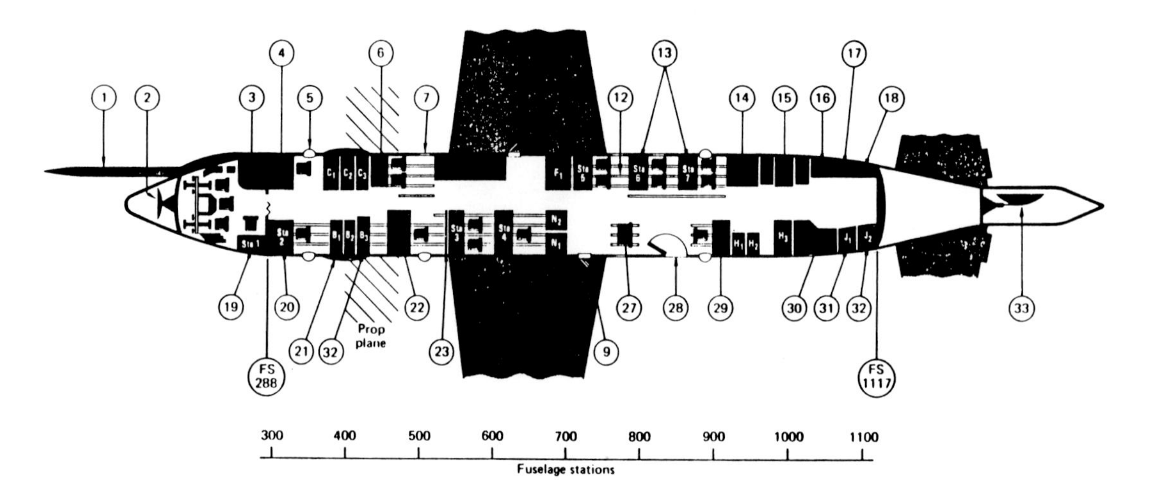

1. Gust probe
2. C–Band radar
3. Fwd elec load center
4. Nav/comm station
5. Bubble window (5 places)
6. Gust probe station
7. Window (6 places)
8. Main elec load center
9. Emergency exits (2)
10. Nav system
11. Dropsonde & AXBT console
12. Floor tracks (typical)
13. Scientific consoles
14. Lavatory
15. Dinette
16. Overhead bunks (2, fold up)
17. Electronic technician work area & storage space
18. Radar recorders
19. Mission scientist station
20. Flight director station
21. C–Band radar
22. Cloud physics station
23. Analog panel
24. Radar console
25. Tape & disk recorders
26. Computer/data system
27. Passenger seats
28. Entry door
29. Camera control/scientific console
30. Galley & galley seats (2)
31. Radar data recording
32. Radar R/T units
33. X–Band radar

NOAA WP-3D interior arrangement. (Lockheed)

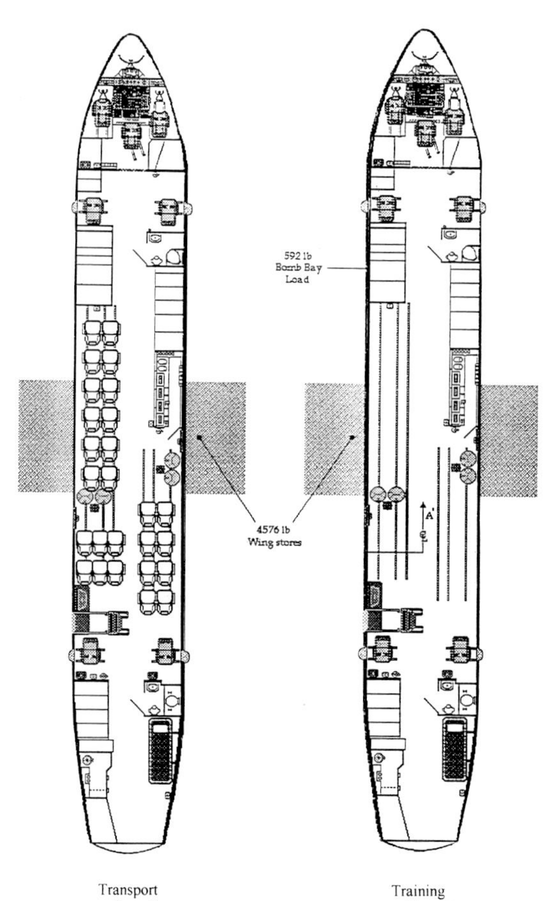

TAP-3 2-plane interior arrangement (ferry/training configurations). (RAAF)

Appendix D
P-3 Orion Bureau Number List

First production YP3V-1 Orion (#148883) ***JEM Aviation***

LASC NO	BuNo	TYPE	LOCATION	COMMENT
9998	xxxxxx	P3V-1	LASC Burbank	(Fatigue Test Article)

*** this was the first P-3 airframe off the production line that never flew, it was held in an iron cage of hydraulic gears - computer controlled to bend, stress and twist the airframe to simulate a lifetime of flight with shear, turbulence, high "G" and landing loads factored in for the P-3 structural test program. The airframe compiled two 7500 hour lifetime cycles (one lifetime completed before the first production flight test aircraft ever rolled out of the plant) - later scrapped and now at Lockheed's old Burbank production plant's boneyard ***

1003 148276 NP-3A MUSEUM

*** was a Lockheed L-188 Electra re-engineered as the YP3V-1 Prototype aircraft for the P-3 Orion VPX program and later the Lockheed/Navy flight test program. The aircraft was later acquired by NASA in 1967 as its NP-3A remote sensing flying laboratory - Now retired by NASA and assigned to the National Museum of Naval Aviation in Pensacola Florida ***

5001 148883 NP-3D USN

*** was first production P3V-1 Orion delivered to the USN and used as a flight test program aircraft - later assigned to the then NADC RDT&E facility as an experimental testbed/prototype aircraft for different Orion programs including A-NEW, SAR, BEARTRAP, INS and more recently the HARPASS reconnaissance system - was re-designated NP-3D in 1994 ***

5002 148884 P-3A STRIKE

*** wing damage during re-work at NADEP Jax on 8/6/77 ***

5003 148885 UP-3A DRMO Jax (for disposal)

*** was SARDIP at the NADEP Jax bone yard and transferred to the local DRMO in 1993 - donated a tail empennage to the NCAR Electra ***

5004 148886 P-3A K-Tech Aviation

*** this is one of eight Orions that AMARC auctioned off in 1990- all of which were acquired by K-Tech Aviation of Tucson, AZ.. Some of the stripped airframes have been sold off to aeronautical companies as R&D mock-up engineering laboratories or avionics integration platforms ***

5005 148887 EP-3E AMARC - FLA

*** was an EP-3E ARIES-Deepwell Orion - recently sent to AMARC pending induction into the EP-3E ARIES II CILOP program as a component donor ***

5006 148888 EP-3E NADEP Alemeda - bone yard

*** was EP-3E ARIES Orion - that was later stripped by NADEP Alameda as a component donor to ARIES II/CILOP program ***

5007 148889 NP-3D USN

*** a versatile testbed aircraft - was once used as the S-3Viking Electronics Flight Test aircraft which consisted of an S-3 flight deck and mission suites installed inside the Orions fuselage. Modified numerous times one of the most unusual was the "Droopsnoot" with a look down radar systems in a 90 degree drooping radome - recently re-designated NP-3D ***

5008 149667 RP-3A STRIKE

*** was the first Project Seascan RP-3A with VXN-8 - later retired to NADEP Jax and donated to NAS Jax emergency crash unit for fire fighting training and burnt up ***

EP-3B "BATRACK" Orion (#149669) ***Isham collection***

Miss Piggy (#149673) ***B. Rogers via Isham collection***

Chilean P-3 (#149677) ; seen here as a logistics / transport aircraft for VQ-2. ***via Baldur Sveinsson***

EP-3B Orion (#149678) ***via Baldur Sveinsson***

5009 149668 EP-3E NADEP Jax (ARIES)

*** was EP-3E ARIES Orion - that was later stripped by NADEP Jax as a component donor for the ARIES II / CILOP program ***

5010 149669 EP-3B LASO AEROMOD

*** was a "Black P-3" used by the CIA in the mid 1960's, later modified as the EP-3B "BAT RACK" prototype for VQ-1, then later brought up to the ARIES configuration and flown until it was retired- scrapped as the first mission systems donor to the ARIES II / CILOP program at Lockheed's AEROMOD facility in Greenville S.C. ***

5011 149670 RP-3A AMARC- FLA

*** was the first Project Birdseye RP-3A with VXN-8 before being provided to NRL as a Oceanographic Research Orion ***

5012 149671 EP-3A USN

*** was a EP-3A weapons test aircraft at PMTC (NAWC Point Mugu) - later retired and acquired as a "static ground airflow generator - wind machine" at the Naval Air Weapons Test Center, China Lake test range ***

5013 149672 P-3A STRIKE

*** water collision at Pax River with VP-8 on 1/30/63 ***

5014 149673 EP-3B AMARC - FMS (Miss Piggy)

*** was one of the first production fleet P-3A delivered to VP-8 in 1962 - later the aircraft was borrowed by the CIA (Black P-3) during the mid 1960's until it was modified as the first EP-3A in 1967 and used as an EW development aircraft with PMTC, MWL and later with VX-1 as the EMPASS BIRD. Re-designated a UP-3A, it served with VP-MAU (NASB) and lastly with VP-30 before retirement to AMARC ***

5015 149674 NP-3D NRL (was a WP-3A)

*** was a P-3A Orion modified by Lockheed into a WP-3A for the US Navy's Weather Reconnaissance Squadron Four (VW-4) "The Hurricane Hunters"- later the aircraft was re-modified into a EP-3A research platform for NRL. Recently re-designated an NP-3D ***

5016 149675 VP-3A ETD- NAS Barbers Point

*** was another Orion modified into a WP-3 - later converted to a VP-3A for VIP Transport ***

5017 149676 VP-3A USN

*** was also a WP-3A that was later re-modified into a VP-3A for VIP Transport ***

5018 149677 P-3A CHILE (#403)

5019 149678 EP-3B LASO AEROMOD

*** was the third P-3A utilized by the CIA in the mid 1960's (Black P-3) - later modified as the second EP-3B "BAT RACK" Orion for VQ-1. Subsequently upgraded to the ARIES configuration, the aircraft was later scrapped - stripped and cutup at AEROMOD as a mission systems donor for the Aires II / CILOP program. The wings from this aircraft were utilized for a joint USN/JMSDF P-3 structural analysis study into fatigue and corrosion problems in developing new preventative measures and procedures ***

5020 150494 EP-3E LASO AEROMOD

*** was an EP-3E ARIES modified Orion - later stripped by AEROMOD as a mission system donor for the ARIES II / CILOP program ***

5021 150495 UP-3A USN (Keflavick, Iceland)

*** a converted P-3A for logistical and passenger transport support to NAS Keflavick ***

5022 150496 VP-3A USN

*** was a WP-3A Orion - later converted to a VP-3A for VIP Transport. The aircraft recently flew to Moscow (3/14/93) and became the first P-3 to land on Russian territory ***

5023 150497 EP-3E NADEP Alameda - boneyard

*** was an EP-3E ARIES-Deepwell modified Orion - later stripped by NADEP Alameda as a mission systems donor to the ARIES II/ CILOP program ***

5024 150498 EP-3E Aero Union (spares aircraft)

*** was an EP-3E ARIES-Deepwell modified Orion - later stripped by NADEP Alameda as a mission systems donor to the ARIES II/ CILOP program - the airframe was recently auctioned off through DRMO to Aero Union. The airframe was parted out and cut up for scrap on site at NADEP Alameda ***

5025 150499 NP-3D USN

*** was first P-3A to be modified with a unique "Billboard" extension of the vertical stabilizer for "EATS" system at the Pacific Missile Test Center - NAWC Point Mugu. Was originally designated EP-3A, then RP-3A and only recently re-designated NP-3D ***

5026 150500 RP-3A TBZ Metals Co.

*** was the second RP-3A "Arctic Fox" Birdseye Orion with VXN-8. The aircraft was later SARDIP at NADEP Jax and turned over to the local DRMO for disposal - disassembled, the aircraft was sold for scrap to the TBZ Metals company of Macon Georgia ***

*** 1976, the aircraft tested a non-acoustic R&D sensor - a crynogentic radiometer probe ***

5027 150501 EP-3E DRMO Alameda

*** was an EP-3E ARIES-Deepwell modified Orion - later stripped by NADEP Alameda as a mission systems donor for the ARIES II / CILOP program ***

5028 150502 EP-3E LASO AEROMOD

*** was an EP-3E ARIES-Deepwell modified Orion - later stripped by AEROMOD as a mission systems donor for the ARIES II/CILOP program ***

5029 150503 EP-3E USN

*** was an EP-3E ARIES-Deepwell modified Orion - later stripped

Base transport UP-3A Orion (#150495) ***B. Tourville via Isham collection***

NAWC Pt. Mugu NP-3D (#150499) ***Via Isham collection***

EP-3E ARIES - Deepwell configured Orion (#150501) ***M. Wada via Baldur Sveinsson***

Partial VIP configured UP-3A Orion (#150504) ***H. Nagakubo via Isham collection***

NAS Moffett Field display aircraft (#150509) ***US Navy***

by AEROMOD as a mission systems donor for the ARIES II/CILOP program, The aircraft was then disassembled with the cockpit section going to Lockheed, to be utilized for the proposed P-7A Flight Simulator (now in storage at Lockheed's Rye Canyon facility) and the aft section of the fuselage sent to China Lake - placed on the weapons test range ***

5030 150504 UP-3A AMARC - FMS

*** a P-3A modified with a unique partial VIP interior configuration once operated by VQ-1 for Commander 7th Fleet (Pacific) ***

5031 150505 EP-3E NADEP Jax (ARIES-Deepwell)

*** was an EP-3E ARIES - Deepwell modified Orion which was later stripped by NADEP Jax as a mission system donor for the ARIES II / CILOP program ***

5032 150506 P-3A K-Tech Aviation

5033 150507 P-3A CHILE (#402)

*** was originally a USN P-3A that was leased to the Spanish Air Force - returned (9/91) to the Navy and retired to AMARC until it was selected by Chile ***

5034 150508 P-3A STRIKE

*** onboard fire, out of Cubi Point the Phillippines, with VP-9 on 12/4/64 ***

5035 150509 P-3A USN (NAS Moffett Field display)

*** was the VP-31 "FRAMP" maintenance training aircraft - recently retired and selected as the Moffett Field display aircraft ***

5036 150510 P-3A Hawkins & Powers Aviation

*** was originally a USN P-3A that was leased to the Spanish Air Force - returned (9/91) to the Navy and retired to AMARC until it was selected by the US Forest Service and provided to H&P airborne fire fighting company of Graybull, WY. - now in GSA Custody ***

5037 150511 VP-3A USN

Surplus P-3A with Hawkins and Powers (#150510) ; seen here as a logistics / transport Orion with VO-2. ***Northolt via Isham collection***

The first USCS P-3A Orion Slick (#150514) ; seen here with VP-94 ***Lockheed***

*** was a P-3A modified into a VP-3A VIP Transport aircraft ***

5038 150512 RP-3A ALCO

*** was a PMTC RP-3A "EATS / SMILS" Orion - recently SARDIP at NADEP Alameda and auctioned off through the local DRMO - it was acquired by the Aircraft Logistical Support Co. of Reno, Nevada - airframe is currently located at American Valley Aviation in Quincy, Ca. (the co-owner) ***

5039 150513 P-3A Texas State Technical College

*** was originally a USN P-3A that was leased to the Spanish Air Force - returned (9/91) to the Navy and retired to AMARC until it was selected by the US Forest Service and provided to Blackhills Aviation (now Neptune Inc.) airborne fire fighting co. - later the aircraft was taken back by the USFS and turned over to GSA custody for disposition. Designated surplus, the aircraft has now been donated to the aviation department of the Texas State Tech. College as a airframe/powerplants training aid used for FAA certification ***

5040 150514 P-3A US Customs Service (N18314)

5041 150515 VP-3A VR-Det/Sigonella

*** was a P-3A modified into a VP-3A VIP Transport aircraft - was flown originally by USMC aircrews for the Marine Commandant (1982-86) - assigned to VP-30's ASA the aircraft remains in Sicily for VIP transport of CinCAFSE ***

5042 150516 P-3A SPAIN (museum)

*** was originally a USN P-3A that was leased to the Spanish Air Force - later retained and purchased (2/15/90) for display in the Spanish Air Force Museum ***

5043 150517 P-3A USN (FRAMP Aircraft)

5044 150518 UP-3A CHILE (#401)

5045 150519 UP-3A AMARC - FMS

*** was once bailed from the USN to the General Offshore Corp. - used for quality assurance testing of sonobuoys for the Navy ***

5046 150520 RP-3A MUSEUM (display)

*** was a weapons test platform with PMTC (NAWC-WD Point Mugu) - later retired and SARDIP by NADEP Alameda - then acquired (1/27/94) by the Western Aerospace Museum of Oakland, Ca. and placed on display ***

5047 150521 NP-3D USN

*** was second "EATS" Billboard modified P-3 Orion - recently re-designated NP-3D***

5048 150522 NP-3D USN

*** was third "EATS" Billboard modified P-3 Orion - recently re-designated NP-3D ***

5049 150523 P-3A Chrysler Technology Inc.

*** auctioned off by the USN to K-Tech Aviation - the aircraft was recently (1994) purchased by Chrysler as an avionics integration

FRAMP P-3 with VP-31 (#150517) ***Terry Taylor***

A range RP-3A with PMTC (#150520) ; the Orion is now a display aircraft at the Western Aerospace Museum in Oakland, California. ***via Isham collection***

A NAWC Pt. Mugu NP-3D (#150525) ; seen here as the PMTC RP-3A "Harpoon" test aircraft. ***M. Grove via Isham collection***

lab ***

5050 150524 NP-3D USN

*** another P-3A modified as an "EATS / SMILS" weapons test platform - has been utilized by the DOE via INFOTEC Development for the "AMPS" program - recently re-designated NP-3D ***

5051 150525 NP-3D USN

*** another P-3A modified as a weapons test platform, configured as the Harpoon missile test aircraft - recently re-designated NP-3D ***

5052 150526 UP-3A USN

*** a P-3A modified with a unique partial VIP interior configuration, currently operated by VQ-1 for COMNAVAIRPAC ***

Once a UP-3A logistical / transport Orion with VXN-8 (#150527) seen here with its "Tasmanian Devil" cartoon. *US Navy*

5053 150527 UP-3A AMARC - FMS (display)

*** was nicknamed "Tasmanian Devil" by VXN-8 and bore that cartoon on the nose -now on AMARC's display row ***

Another UP-3A logistical / transport Orion with VXN-8 (#150528) ; seen here with its "Loon" cartoon. *US Navy*

5054 150528 UP-3A AMARC - FMS

*** was nicknamed "Loon" by VXN-8 and bore that cartoon on the nose ***

5055 150529 EP-3E Hawkins & Powers Aviation

*** a P-3A modified for electronic warfare - counter measures with VAQ-33, later retired by the Navy to AMARC until it was selected by the US Forest Service and provided to H&P airborne fire fighting co. - was recently taken back by the USFS and turned over to GSA custody for disposition ***

5056 150604 P-3A AMARC - REC

5057 150605 UP-3A USN

*** a P-3A recently modified with a unique partial VIP interior configuration as the new backup VIP aircraft for NAS Barbers Point ETD's VP-3A 149675 ***

5058 150606 P-3A K-Tech Aviation

5059 150607 UP-3A CHILE (#406)

*** was operated by both VXN-8 and NRL at one time ***

A fleet P-3A (#150608) ; seen here with VP-66 *B. Stewart*

Once a training / Auxiliary mission P-3A with VW-4 (#151352) ; now a TP-3A. ***via Isham collection***

The first Chilean UP-3A (#151354) ***Bob Shane***

An Aero Union P-3A air tanker #24 (#151359) ; seen here approximately twenty-four hours before it crashed. ***M. Peltzer***

An Aero Union P-3A Aerostar #25 (#151361) ***M. Peltzer***

5060 150608 P-3A AMARC - REC

5061 150609 P-3A K-Tech Aviation

5062 151349 P-3A K-Tech Aviation

5063 151350 P-3A STRIKE

*** water collision in south china sea, with VP-6 on 4/5/68 ***

5064 151351 P-3A E-Systems

*** auctioned off by the USN to K-Tech Aviation - the aircraft was recently (1994) purchased by E-Systems of Greenville, Texas to be used as a test article for the USN SRP program ***

5065 151352 TP-3A USN

5066 151353 UP-3A AMARC - FMS

5067 151354 UP-3A CHILE (#405)

*** was the first aircraft delivered to Chile (3/3/93) ***

5068 151355 P-3A Aero Union

*** retired by the Navy to AMARC until it was selected by the US Forest Service and provided to Aero Union airborne fire fighting co. as a parts aircraft - was recently taken back by the USFS and turned over to GSA custody for disposition ***

5069 151356 P-3A AMARC - FMS

5070 151357 TP-3A USN

5071 151358 UP-3A AMARC - REC

5072 151359 P-3A Aero Union

*** retired by the Navy to AMARC until it was selected by the US Forest Service and provided to Aero Union airborne fire fighting co. and converted into an AEROSTAR fire fighter - tanker #24 (N924AU) - which later crashed into a Montana MT. on Oct. 6th, 1991 ***

5073 151360 P-3A AMARC - FMS

5074 151361 P-3A Aero Union

*** retired by the Navy to AMARC until it was selected by the US Forest Service and provided to Aero Union airborne fire fighting co. and converted into an AEROSTAR fire fighter- tanker #25 (N925AU) - was the first P-3 Orion modified as an air tanker ***

5075 151362 P-3A STRIKE

*** water collision off of Argentia, Newfoundland, with VP-45 on 11/17/64 ***

5076 151363 P-3A STRIKE

*** aborted takeoff, ran off runway at Adak, Alaska, with VP-45 on 6/2/69 ***

5077 151364 TP-3A AMARC - FLA

5078 151365 P-3A STRIKE

A fleet P-3A (#151363) ; seen here after it crashed at Adak Alaska. ***via Harry Repine Jr.***

A specialized UP-3A base transport aircraft for NAS Bermuda (#151367) ; the aircraft Orion has recently become a display at NAS Keflavik, Iceland. ***US Navy***

An Aero Union Aerostar air tanker #27 (#151369) ***Aero Union Corp.***

*** water collision off Japan, with VP-4 on 4/28/67 ***

5079 151366 P-3A GREECE

*** retired by the Navy to AMARC until it was selected by Greece for the Hellenic Navy - the aircraft has since been delivered to Hellenic Aerospace Co. In Greece for spare parts ***

5080 151367 UP-3A USN (Bermuda bird)

*** originally modified into the 12th TP-3A by NADEP Jax, the aircraft later received the first UP-3A conversion mod, becoming the UP-3 Prototype and un-officially designated *UP-3S* for Standardized Utility P-3 Configuration - the aircraft currently is designated as a UP-3A (but with the TP-3A cockpit configuration) assigned to NAS Bermuda as the base transport aircraft - the aircraft has now been transferred to NAS Keflavik as the base display aircraft ***

5081 151368 UP-3A AMARC - FMS

5082 151369 P-3A Aero Union

*** retired by the Navy to AMARC until it was selected by the US Forest Service and provided to Aero Union airborne fire fighting co and converted into an AEROSTAR fire fighter - tanker #27 (N927AU) ***

5083 151370 TP-3A AMARC - FLA

5084 151371 TP-3A USN

5085 151372 P-3A Aero Union

*** retired by the Navy to AMARC until it was selected by the US Forest Service and provided to Aero Union airborne fire fighting co and converted into an AEROSTAR fire fighter - tanker #23 (N923AU) ***

5086 151373 P-3A NADEP Alameda (bone yard)

5087 151374 P-3A USN (NAS Jacksonville display)

*** retired by the Navy and placed on display at the NAS Jacksonville gate ***

5088 151375 TP-3A AMARC - FLA

5089 151376 TP-3A AMARC - FLA

5090 151377 P-3A Aero Union

*** retired by the Navy to AMARC until it was selected by the US Forest Service and provided to Aero Union airborne fire fighting co. as a parts aircraft - was recently taken back by the USFS and turned over to GSA custody for disposition - the aircraft has since been auctioned off and sold back to Aero Union ***

5091 151378 P-3A AMARC - FMS

5092 151379 TP-3A USN

5093 151380 P-3A STRIKE

*** ground collision in Bermuda, with VP-16 on 7/27/65 ***

5094 151381 P-3A STRIKE

*** wheels up landing at NAS Jacksonville, with VP-62 on 2/23/78

Was a TP-3A Orion (#151382) ; seen here as a fleet P-3A with VP-22. ***N.M. Williams via Baldur Sveinsson***

Was briefly a P-3A air tanker, #21 with Aero Union (#151385) ***Knowles via Isham collection***

A fleet P-3A (#151388) ; seen here in rare old VP-91 markings. ***Lockheed***

***

5095	151382	TP-3A	AMARC - FLA	
5096	151383	P-3A	AMARC - FMS	
5097	151384	UP-3A	CHILE	(#407)
5098	151385	P-3A	Aero Union	

*** retired by the Navy to AMARC until it was selected by the US Forest Service and provided to Aero Union airborne fire fighting co and was originally converted to an AEROSTAR fire fighter - tanker #21 (N921AU) - later the aircraft was down graded to a non-fly status and taken back by the USFS and turned over to GSA custody for disposition - the aircraft has since been auctioned off and sold back to Aero Union ***

5099	151386	P-3A	AMARC - REC
5100	151387	P-3A	Aero Union

*** retired by the Navy to AMARC until it was selected by the US Forest Service and provided to Aero Union airborne fire fighting co. and converted in to an AEROSTAR fire fighter - tanker #22 (N922AU) ***

5101	151388	P-3A	AMARC - REC
5102	151389	P-3A	GREECE

*** retired by the Navy to AMARC until it was selected by Greece for the Hellenic Navy - has since been delivered to Hellenic Aerospace Co. For spare parts ***

5103	151390	P-3A	US Customs Service

*** retired by the Navy to AMARC until it was selected by the US Customs Service as a long range intercept aircraft used in counter narcotics operations - known as a SLICK (N15390) ***

5104	151391	P-3A	Aero Union

*** retired by the Navy to AMARC until it was selected by the US Forest Service and provided to Aero Union airborne fire fighting co. and converted into an AEROSTAR fire fighter - tanker #00 (N900AU) - in 1991 the aircraft was used as an engine testbed aircraft for GMA-Allison's GMA 2100 (406) turboprop engine (the aircraft bore the registration of N406TP) ***

5105	151392	TP-3A	USN
5106	151393	P-3A	AMARC - REC
5107	151394	TP-3A	AMARC - FLA
5108	151395	P-3A	US Customs Service

*** retired by the Navy to AMARC until it was selected by the US Customs Service as a long range intercept aircraft used in counter narcotics operations - known as a SLICK (N16295) ***

5109	151396	P-3A	AMARC - REC
5110	152140	P-3A	AMARC - FMS (TNM)

An Aero Union air tanker #00 (#151391) ; the aircraft is seen here as the Allison GMA- 2100 turboprop engine testbed. ***Allison***

A fleet P-3A (#151396) ; seen here as a reclamated hulk at AMARC. ***B. Stewart***

A display aircraft at the Pensacola Museum (#152152) ; seen here as a fleet P-3A with VP-6. ***via Baldur Sveinsson***

5111 152141 UP-3A CHILE (#408)

5112 152142 P-3T THAILAND

5113 152143 P-3T THAILAND

5114 152144 P-3A STRIKE

*** collision with mountain in Japan, with VP-48 on 1/16/68 ***

5115 152145 P-3A SPAIN (#22-22 / P.3-3)

5116 152146 P-3A AMARC - FMS (TNM)

5117 152147 P-3A AMARC - FMS

*** was once bailed to the General Offshore Corporation that had a Navy contract to provide sonobuoy quality assurance testing ***

5118 152148 P-3A AMARC - FMS (TNM)

5119 152149 P-3A STRIKE

*** crashed with the Spanish Air Force in Spain on 7/8/77 ***

5120 152150 NP-3D USN

*** a versatile testbed aircraft for NADC - specially configured with four optical windows in the floor (26 sq.ft. Of observing area) to house various sensors, cameras and lasers- used for the ASW Laser research project and served as the *LIDAR* laser testbed aircraft, earning the nickname " Glass Bottom Orion"... was later utilized for the *ERAPS* ASW project aircraft and was used to test the durability of the *UNICOAT* anti-corrosive paint coating with the port side of the Orion painted with the experimental coating and the other side painted with standard primer and paint - originally designated a UP-3A upon leaving the fleet and assuming its career as a

A fleet P-3A that disappeared with no trace (#152155) ***US Navy***

testbed, it was only recently re-designated an NP-3D ***

5121 152151 P-3A STRIKE

*** engine failure (cleaning solvent) in Cubi Point, the Phillippines, with VP-6 on 12/5/71***

5122 152152 P-3A MUSEUM (display)

*** on display at the National Museum of Naval Aviation, Pensacola, Florida ***

5123 152153 P-3A SPAIN (#22-21 / P.3-1)

5124 152154 P-3A AMARC - FMS (TNM)

5125 152155 P-3A STRIKE

*** disappeared off California coast, with VP-31 on 5/26/72 ***

5126 152156 P-3A USN (NAS Brunswick display)

5127 152157 P-3A AMARC - FMS (TNM)

5128 152158 P-3A USN (NAWC -23)

5129 152159 P-3A STRIKE

*** mid-air explosion (possible lightning strike) over Nevada, with VP-17 on 8/3/70 ***

5130 152160 P-3A Bermuda (ex-display)

*** has been the gate guard display aircraft since 1990 and now turned over to the Bermuda government - was to have been sunk as a shallow water dive site, but has now been burnt up for local airport fire fighting training and disposed of on site ***

A display aircraft at NAS Brunswick, Maine (#152156) ***P-3 Publications***

A fleet P-3A (#152158) ; seen here in VP-64 markings. ***D. Brown***

5131 152161 P-3A STRIKE

*** hard landing at NAS Whidbey Island, with VP-69 on 1/18/81 ***

5132 152162 P-3A AMARC - FMS (TNM)

5133 152163 P-3A THAILAND

*** delivered to Thailand 1/94 as a parts aircraft and ground (maint.) trainer ***

5134 152164 P-3A K-Tech Aviation

*** retied by the Navy at NADEP Alameda, it was SARDIP and turned over to the local DRMO for disposition - auctioned off to K-Tech for remaining parts and cut up on site at Alameda (provided tail empennage to US Customs SLICK 152170) ***

5135 152165 P-3A CHILE (#404)

5136 152166 P-3A STRIKE

*** hard landing at NAS Whidbey Island, with VP-69 on 1/ ?/89 ***

5137 152167 P-3A AMARC - FMS (TNM)

5138 152168 P-3A AMARC - FMS (TNM)

5139 152169 UP-3A USN

5140 152170 P-3A US Customs Service

*** retired by the Navy to AMARC until it was selected by the US Customs Service as a long range intercept aircraft used in counter narcotics operations - known as a SLICK (N16370) ***

5141 152171 P-3A STRIKE

*** water collision off Norris, California with VP-19 on 4/9/66 ***

5142 152172 P-3A STRIKE

*** ground collision in Michigan, with VP-19 on 7/4/66 ***

5143 152173 P-3A AMARC - FMS (TNM)

5144 152174 P-3A AMARC - FMS (TNM)

5145 152175 P-3A AMARC - FMS (TNM)

5146 152176 P-3A AMARC - FMS (TNM)

5147 152177 P-3A THAILAND

*** delivered to Thailand 1/94 as parts aircraft and ground (maint) trainer ***

5148 152178 P-3A K-Tech Aviation

5149 152179 UP-3A AMARC - FMS

5150 152180 P-3A AMARC - FMS (TNM)

Was the display aircraft at NAS Bermuda (#152160) *P-3 Publications*

A fleet P-3A later retired and sold off as scrap (#152164); seen here in VP-69 markings. *D. Brown*

A fleet P-3A stored with others at AMARC (#152167) *B. Stewart*

A surplus fleet P-3A sold to Greece (#152181) ; seen here in VP-60 markings. *B. Stewart via Baldur Sveinsson*

A fleet P-3A stored with others at AMARC (#152186) ***B. Stewart***

The first EP-3J Orion C³I /CM aircraft (#152719) ***Terry Taylor***

5151 152181 P-3A GREECE

*** retired by the Navy to AMARC until it was selected by Greece for the Hellenic Navy - the aircraft has since been delivered to Hellenic Aerospace Co. As a ground maintenance trainer ***

5152 152182 P-3A STRIKE

*** ground collision (with mountain) in Morocco, with VP-44 on 6/3/72 ***

5153 152183 P-3A GREECE

*** retired by the Navy to AMARC until it was selected by Greece for the Hellenic Navy - the aircraft has been since been delivered to Hellenic Aerospace Co. As ground maintenance trainer ***

5154 152184 UP-3T THAILAND

5155 152185 P-3A AMARC - FMS

5156 152186 P-3A AMARC - FMS (TNM)

5157 152187 P-3A AMARC - FMS (TNM)

5158 152718 P-3B AMARC - HOLD (LW/TNM)

5159 152719 EP-3J USN (C³I / CM aircraft)
*** produced as the first fleet LW Bravo, the aircraft became one of two modified with a unique comand, control, communications and intelligence/counter-measures capability for EW training to fleet battle groups at sea and re-designated EP-3J (the J for Jamming) - The EP-3J operated first with VAQ-33 out of NAS Key West, Florida but were later transferred to reserve squadron VP-66 at NAS Willow Grove when VAQ-33 was retired in 1993 - now assigned to the new VQ-11 Squadron ***

5160 152720 P-3B STRIKE

*** ground collision with mountain in Hawaii with VP-1 on 6/16/83 ***

The third P-3 AEW&C for US Customs (#152722) ***Lockheed***

A specialized NP-3B Orion (#152739) ; seen here with a large, elongated ventral canoe pod. ***B. Neidermer via Isham collection***

5161 152721 P-3B AMARC - FMS (LW/TNM)

5162 152722 P-3 AEW&C US Customs Service

*** retired by the Navy and provided directly to the US Customs Service as a baseline aircraft for modification (performed by Lockheed) into the third P-3 airborne early warning and control P-3 or P-3 AEW&C used in counter narcotics operations - known as a DOME (N147CS) ***

5163 152723 P-3B AMARC - FMS (LW/TNM)

5164 152724 P-3B STRIKE

*** water collision off the Azores with VP-23 on 4/26/78 ***

5165 152725 P-3B AMARC - FMS (LW/TNM)

5166 152726 P-3B AMARC - FMS (LW/TNM)

5167 152727 UP-3B AMARC -FLA

5168 152728 P-3B USN (specially equipped)

5169 152729 P-3B AMARC - FMS (LW/TNM)

5170 152730 P-3B AMARC - FMS (LW/TNM)

5171 152731 P-3B NOAA (spares aircraft)

*** retired by the Navy and provided directly to NOAA as a spare parts aircraft with potential as a project aircraft - in storage at NOAA's aircraft operations center at MacDill AFB Tampa, Florida ***

5172 152732 P-3B AMARC - FMS (LW/TNM)

5173 152733 P-3B STRIKE

*** wheels up landing in Hawaii with VP-1 on 5/17/83 ***

5174 152734 P-3B AMARC - FMS (LW/TNM)

5175 152735 P-3B NASA (EFIS)

*** retired by the Navy and provided directly to NASA and modified into a remote sensing aircraft for NASA and other science user research projects - equipped with a "Glass Cockpit" or electronic flight instrument system (EFIS) - (N426NA) ***

5176 152736 P-3B AMARC - FMS (LW/TNM)

5177 152737 P-3B AMARC - FMS (LW/TNM)

5178 152738 RP-3D AMARC - FMS (Mini-mod)

*** produced as a fleet LW Bravo, the aircraft later received "mini-mods"giving it a roll-on capability for the Projects Birdseye and Seascan with VXN-8 - the aircraft was retired to AMARC in 1993 when VXN-8 disestablished ***

5179 152739 NP-3B USN (NAWC - 23)

*** produced as a fleet LW Bravo that was later acquired by NAWC and modified into a EW R&D projects aircraft - most notable is a elongated ventral canoe pod stretching from the weapons bay back to the main cabin door ***

5180 152740 UP-3B AMARC - FMS

5181 152741 P-3B AMARC - FMS (LW/TNM)

5182 152742 P-3B AMARC - FMS (LW/TNM)

5183 152743 P-3B AMARC - FMS (LW/TNM)

5184 152744 P-3B GREECE

*** produced as a fleet LW Bravo and later retired by the Navy to AMARC until it was selected by Greece for the Hellenic Navy ***

A fleet P-3B (#152751) ; seen here in VP-60 markings. *P. Minery via Baldur Sveinsson*

A fleet P-3B lost over Maine (#152757) ; seen here in old VP-31 markings. *US Navy*

A fleet P-3B stored with others at AMARC (#152757) *Bob Shane*

5185 152745 EP-3J USN (C³I / CM aircraft)

*** produced as a fleet LW Bravo, was modified with a unique command, control, communications and intelligence/counter-measure capability for EW training of fleet battle groups at sea and re-designated EP-3J (the J for Jamming) - the aircraft first operated with VAQ-33 out of NAS Key West, Florida but was later transferred reserve squadron VP-66 when VAQ-33 retired in 1993 - now assigned to VQ-11 ***

5186 152746 P-3B AMARC - FMS (LW/TNM)

5187 152747 P-3B GREECE (option aircraft)

*** produced as a fleet LW Bravo and later retired by the Navy to AMARC until it was selected by Greece for the Hellenic Navy as an operational aircraft ***

5188 152748 P-3B USN (Selfridge NARA display)

*** retired by the Navy and placed on display at the old NAS Detroit, Michigan base ***

5189 152749 P-3B STRIKE

*** water collision off of Brunswick, Maine with VP-10 on 3/15/73 ***

5191 152750 P-3B AMARC - FMS (LW/TNM)

5193 152751 P-3B AMARC - FMS (LW/TNM)

5194 152752 P-3B AMARC - FMS (LW/TNM)

5195 152753 P-3B AMARC - FMS (LW/TNM)

5196 152754 P-3B AMARC - FMS (LW/TNM)

5197 152755 UP-3B AMARC - FMS (LW/TNM)

5198 152756 P-3B AMARC - FMS (LW/TNM)

5199 152757 P-3B STRIKE

*** wing separation over Maine with VP-8 on 10/22/78 ***

5201 152758 P-3B AUSTRALIA (TAP-3 aircraft)

*** retired by the Navy to AMARC until it was selected by Australia for its RAAF TAP-3 Project to convert three LW Bravos to training / logisitcs aircraft - the MOD is being performed by NADEP Jacksonville ***

*** this aircraft was once operated by NATC at NAS Pax River as a systems testbed used for such projects as the feasibility tests of P-3 in-flight re-fueling systems ***

5203 152759 P-3B AMARC - FMS (LW/TNM)

5204 152760 P-3B AUSTRALIA

*** retired by the Navy to AMARC, this aircraft has now been selected by Australia as spare parts aircraft for its RAAF TAP-3 Project - the aircraft fuselage is also expected to be utilized as a test article for a battle damage repair study and AWAA OMS cockpit simulator ***

A New Zealand P-3K (#152889) *Boeing*

A fleet P-3B stored with others at AMARC (#153418) *B. Stewart*

5205	152761 P-3B	AMARC - FMS (LW/TNM)	
5206	152762 P-3B	AMARC - FMS (LW/TNM)	
5207	152763 P-3B	AMARC - FMS (LW/TNM)	
5209	152764 P-3B	USN (NAS Whidbey Is. display)	
5210	152765 P-3B	STRIKE	

*** hard landing/fire at NAS Lemoore (Calif.) with VP-31 on 3/6/69 ***

5190	152886 P-3K	NEW ZEALAND (#NZ4201)	

*** produced by Lockheed as export LW (Deltic) Bravos for New Zealand, the aircraft were later uniquely upgraded by Boeing and re-designated P-3K (the K for Kiwi) ***

5192	152887 P-3K	NEW ZEALAND (#NZ4202)	
5200	152888 P-3K	NEW ZEALAND (#NZ4203)	
5202	152889 P-3K	NEW ZEALAND (#NZ4204)	
5208	152890 P-3K	NEW ZEALAND (#NZ4205)	
5211	153414 P-3B	AMARC -MUSEUM	

*** now retired by the Navy to AMARC and assigned to the National Museum of Naval Aviation in Pensacola, Florida and designated a flying trader ***

5212	153415 P-3B	GREECE	(option aircraft)

*** produced as a fleet LW Bravo and later retired by the Navy to AMARC until it was selected by Greece for the Hellenic Navy as an operational aircraft ***
*** this aircraft, while with the reserves, was used as a testbed aircraft for IPADS and modified with a P-3C update III like acoustic processor and display system ***

A fleet P-3B acquired by NASA as a parts aircraft (#153429) *NASA*

5213 153416 P-3B AMARC - FMS (LW/TNM)

5214 153417 P-3B AMARC - FMS (LW/TNM)

5215 153418 P-3B AMARC - FMS (LW/TNM)

5216 153419 P-3B AMARC - FMS (LW/TNM)

5217 153420 P-3B AMARC - FMS (LW/TNM)

5218 153421 P-3B AMARC - FMS (LW/TNM)

*** this aircraft was once used by VX-1 to test the British SEARCHWATER Radar System on a P-3 (between 1982 -83) ***

5219 153422 P-3B AMARC - FMS (LW/TNM)

5220 153423 P-3B AMARC - FMS (LW/TNM)

5221 153424 P-3B GREECE

*** produced as a fleet LW Bravo and later retired by the Navy to AMARC until it was selected by Greece for the Hellenic Navy as an operational aircraft ***

5222 153425 UP-3B STRIKE (SARDIP)
*** produced as a fleet LW Bravo, the aircraft was later re-configured as a UP-3B for used by VQ-2 - later the aircraft developed severe corrosion and was SARDIP at OGMA (in Portugal) and cut up and disposed of on-site ***

5223 153426 P-3B AMARC - FMS (LW/TNM)

*** once had a mid-air collision with a USCG C-130 over Midway Island (on 12/12/71) and had eight feet of wing torn off - the aircraft was safely landed and later repaired to resume duites until it was retired by the Navy to AMARC ***

5224 153427 P-3B GREECE

*** produced as a fleet LW Bravo and later retired by the Navy to AMARC until it was selected by Greece for the Hellenic Navy as an operational aircraft ***

5225 153428 P-3B STRIKE

*** ground collision with a mountain in the Canary Islands with VP-11 on 12/11/77 ***

5226 153429 P-3B AMARC - FLA (NASA)

*** produced as a fleet LW Bravo, the aircraft was later retired by the Navy and provided directly to NASA as a spare parts aircraft - that was then flown by NASA to AMARC and placed in storage ***

5227 153430 P-3B AMARC - FMS (LW/TNM)

5228 153431 P-3B AMARC - FMS (LW/TNM)

5229 153432 P-3B AMARC - FMS (LW/TNM)

5230 153433 UP-3B USN

5231 153434 P-3B AUSTRALIA (TAP-3)
*** produced as a fleet LW Bravo and later retired by the Navy to AMARC until it was selected by Australia for its RAAF TAP-3 Project ***

A fleet P-3B in storage at AMARC (#153434) ***B. Stewart***

A Norwegian P-3N (#154576) ***RNoAF***

A dual-mission capable NP-3D with NRL (#154587) ***Terry Taylor***

5232 153435 P-3B AMARC - FMS (LW/TNM)

5233 153436 P-3B AMARC - FMS (LW/TNM)

5234 153437 P-3B AMARC - FMS (LW/TNM)

5235 152438 P-3B AMARC - FMS (LW/TNM)

5236 153439 P-3B AUSTRALIA (TAP-3)

*** produced as a fleet LW Bravo and later retired by the Navy to AMARC until it was selected by Australia for its RAAF TAP-3 Project ***

5237 153440 P-3B STRIKE

*** water collision in the South China Sea (after enemy action) with VP-26 on 2/6/68 ***

5238 153441 P-3B GREECE

*** produced as a fleet LW Bravo and later retired by the Navy to AMARC until it was selected by Greece for the Hellenic Navy as an operational aircraft - was first mission aircraft delivered to Greece (5/22/96) ***

5239 153442 NP-3D NRL

*** produced as the first fleet Heavy Weight Bravo, it was later modified as a C I/CM EW simulator-evaluator aircraft for NRL and re-designated EP-3B (electronic testbed aircraft) - was recently re-designated an NP-3D ***

5500 153443 NP-3D USN

*** produced as a fleet HW Bravo, the aircraft was later re-engineered by Lockheed as the YP-3C prototype for the P-3C program until later when it was modified again into a Dual- Mission RP-3D airborne research aircraft by NADEP Jax with the capability to conduct both the Seascan and Birdseye Projects for VXN-8. The aircraft was transferred to NAWC Pax River when VXN-8 retired in 1993 - was recently re-designated an NP-3D ***

5240 153444 P-3B USN (NAS New Orleans display)

*** was once involved in the Myagues incident with the aircraft taking hits from enemy fire while in direct communications with the White House ***

5241 153445 P-3B STRIKE

*** water collision in the South China Sea as a result of direct enemy action, with VP-26 on 4/1/68 ***

5242 153446 P-3B AMARC - FLA (HW/TNM)

5243 153447 P-3B AMARC - FLA (HW/TNM)

5244 153448 P-3B AMARC - FLA (HW/TNM)

5245 153449 P-3B AMARC - FLA (HW/TNM)

5246 153450 P-3B USN

*** produced as a HW Bravo later modified with a DIFAR acoustics capability until being re-engineered as a special projects Orion with VPU-1 ***

5247 153451 P-3B AMARC - FLA (HW/TNM)

5248 153452 P-3B AMARC - FLA (HW/TNM)

5249 153453 P-3B AMARC - FLA (HW/TNM)

5250 153454 P-3B K-Tech Aviation

*** retired by the Navy at NADEP Alameda, it was SARDIP and turned over to the local DRMO for disposition - were it was auctioned off to K-Tech for remaining parts and cut up on site — donated wing to P-3B 154599 ***

5251 153455 P-3B AMARC - FLA (HW/TNM)

5252 153456 P-3B AMARC - FLA (HW/TNM)

5253 153457 P-3B AMARC - FLA (HW/TNM)

5254 153458 P-3B AMARC - FLA (HW/TNM)

5255 154574 P-3B USN (NAS Willow Grove display)

5256 154575 P-3 AEW&C US Customs Service

*** retired by the Navy and provided directly to the US Customs Service as a baseline aircraft for modification (performed by Lockheed) into the forth P-3 airborne early warning and control or P-3 AEW&C used in counter narcotics operations - known as a DOME (#N148CS) ***

5257 154576 P-3N NORWAY (#4576)

*** produced as a fleet HW Bravo, the aircraft was later sold to the RNoAF - recently the aircraft was modified by NADEP Jax for coast guard duties/special missions and re-designated P-3N ***

5258 154577 P-3B USN

*** produced as a fleet HW Bravo later modified with a DIFAR acoustics capability until being re-engineered as a special projects Orion with VPU-1 ***

5259 154578 P-3B AMARC - FLA (HW/TNM)

5260 154579 P-3B AMARC - FLA (HW/TNM)

5261 154580 P-3B AMARC - FLA (HW/TNM)

5262 154581 P-3B AMARC - FLA (HW/TNM)

5263 154582 P-3B AMARC - FLA (HW/TNM)

5264 154583 P-3B SPAIN (#22-31 / P.3-8)

*** produced as a fleet HW Bravo, the aircraft was later sold to Norway - who years later sold it to Spain, where it is now in the service of the Spanish Air Force ***

A multi-mission airborne research platform NP-3D with NRL (#154589) ***Terry Taylor***

5265 154584 P-3B AMARC - FLA

*** produced as a fleet HW Bravo later modified with a DIFAR acoustic capability until being re-modified for use by VPU-1 - now retired by the Navy to AMARC ***

5266 154585 P-3B NADEP Jax (SARDIP)

*** produced as a fleet HW Bravo later modified with a DIFAR acoustics capability until being re-modified for use by VPU-2 - now retired by the Navy to NADEP Jax and SARDIP - recently donated its epennage tail section to P-3C 161009 ***

5267 154586 P-3B AMARC - FLA (HW/TNM)

5268 154587 NP-3D NRL (Arctic Fox)

*** produced as a fleet HW Bravo later modified into a Dual-Mission RP-3D airborne research aircraft by NADEP Jax with the capability to conduct both the Seascan and Birdseye Projects for VXN-8. The aircraft was transferred to NAWC Pax River when VXN-8 retired in 1993 - was recently re-designated an NP-3D ***

5269 154588 P-3B AMARC - FLA (HW/TNM)

*** produced as a fleet HW Bravo, the aircraft was later retired by the reserves and transferred to NAWC-WD Point Mugu as an interim range aircraft ***

5270 154589 NP-3D NRL

*** produced as a fleet HW Bravo later modified into a multi-purpose airborne research aircraft by NADEP Jax designed with a back-up capability for the "EW" mission of 153442 and tested the LADS system (1995-96) - was recently re-designated an NP-3D ***

5271 154590 P-3B AMARC - FLA (HW/TNM)

5272 154591 P-3B STRIKE

*** wheels up landing in Hawaii with VP-6 on 9/5/80 ***

5273 154592 P-3B AMARC - FLA (HW/TNM)

5274 154593 P-3B AMARC - FLA (HW/TNM)

5275 154594 P-3B AMARC - FLA (HW/TNM)

5276 154595 P-3B AMARC - FLA (HW/TNM)

5277 154596 P-3B STRIKE

*** engine fail /fire at Cubi Point with VP-22 on 6/27/79 ***

5278 154597 P-3B AMARC - FLA (HW/TNM)

5279 154598 P-3B AMARC - FLA (HW/TNM)

5280 154599 P-3B AMARC - FLA (HW/TNM)

*** the aircraft received a replacement wing on 6/91 from 153454 donor Orion ***

5281 154600 RP-3D AMARC - FLA

*** produced as a fleet HW Bravo, the aircraft later received "minimods" giving it a roll-on capability for projects Birdseye and Seascan with VXN-8 - The aircraft was retired to AMARC in 1993 when VXN-8 disestablished ***

5282 154601 P-3B AMARC - FLA (HW/TNM)

5283 154602 P-3B AMARC - FLA (HW/TNM)

5284 154603 P-3B AMARC - FLA (HW/TNM)

5285 154604 P-3B AMARC - FLA (HW/TNM)

5286 154605 P-3 AEW&C US Customs Service

*** produced as a fleet HW Bravo for the Navy, the aircraft was later provided to Australia via Lockheed as a replacement for RAAF P-3B 155296 that was destroyed by fire (crash landing) before delivery to the RAAF in 1968 - the aircraft was later sold back to Lockheed (traded in towards the purchase of new P-3C) and then re-engineered by Lockheed as the second P-3 airborne early warning and control aircraft for the US Customs Service used in counter narcotics operation - known as a DOME (N146CS) ***

5401	155291 P-3K	NEW ZEALAND (#NZ4206)

*** produced as the first of ten DELTIC HW Bravo sold to Australia's RAAF - the aircraft was later re-sold to New Zealand by Australia in 1985 and was subsequently modified as an P-3K by Boeing ***

5402	155292 P-3P	PORTUGAL	(#4801)

*** produced as a DELTIC HW Bravo for Australia, the aircraft (one of six) was later sold back to Lockheed towards the purchased of new P-3C and later re-engineered by Lockheed, the prototype P-3P Orion for Portugal ***

A Portuguese P-3P Orion (#155292) ; seen here in original paint scheme and markings. *PoAF*

5403	155293 P-3P	PORTUGAL	(#4802)
5404	155294 P-3P	PORTUGAL	(#4803)
5405	155295 P-3P	PORTUGAL	(#4804)
5406	155296 P-3B	STRIKE	

*** landing gear mishap /fire at Moffett Field with the RAAF on 4/11/68 ***

*** produced as an Australian P-3B (A9-296) - just two weeks out of the factory and undergoing acceptance trials prior to delivery to Australia, the aircraft crash landed at NAS Moffett Field on February 11, 1968 - the incident happened after a series of touch and goes where the left main wheels and axle broke off from the main gear leg- upon touch down the aircraft began to ground looped as the propellers disintegrated with #2 engine prop smashing through the fuselage, cutting the radar operators station in half before the plane burst into flames (a subsequent investigation revealed that gear failure was due to a fault in the manufacture of the undercarriage leg forgings) and was completely destroyed - after the incident negotiations between the RAAF and Lockheed resulting in a replacement P-3B being arranged and delivered to the RAAF from the US Navy - #154605 / A9-296 ***

A Portuguese P-3P Orion (#155295) ; close-up view. *via Marco Borst*

5407	155297 P-3P	PORTUGAL	(#4805)
5408	155298 P-3P	PORTUGAL	(#4806)
5409	155299 P-3	AEW&C	US Customs Service

*** produced as a DELTIC HW Bravo for Australia, the aircraft was later sold back to Lockheed (towards the purchase of new P-3C) and was re-engineered by Lockheed as the prototype P-3 airborne early warning and control aircraft for US Customs Service used in counter narcotic operations - known as a DOME (N145CS) ***

5410	155300 ORACL	AUSTRALIA	(old 300)

*** produced as a DELTIC HW Bravo for Australia, the aircraft was to have been sold back to Lockheed (towards the purchase of new P-3C) but became the victim of an oxygen fire on 1/27/84 during ground maintenance work at RAAF Base Edinburgh with the cockpit completely burnt and damaged beyond repair - later the aircraft supplied spare parts to Portuguese P-3P and donated a wing to a US Navy P-3C (157330) that had crash landed (wheels up) during February 1988 In 1992, Australia's Defense and Technology Orga-

A P-3B Orion (155296) ; seen here as a Australian Bravo that crashed and burned at NAS Moffett Field. *US Navy*

nization - Aeronautical Research Laboratory refurbished the hulk airframe as a mock-up demonstrator/simulator used in the development of the RAAF P-3C Refurbishment / AP-3C program ***

5501	156507	EP-3E	USN	(ARIES II)

*** produced as a fleet P-3C, the aircraft was later re-engineered (one of twelve) by Lockheed into the EP-3E ARIES II prototype aircraft ***

5502	156508	P-3C	USN	NUD

*** produced as a fleet P-3C, the aircraft is to be destroyed as part of the SRP test program ***

5503	156509	P-3C	USN	NUD
5504	156510	P-3C	USN	Update III MOD
5505	156511	EP-3E	USN	(ARIES II)
5506	156512	P-3C	AMARC - FLA	NUD

*** aircraft once set a time/distance record from Atsugi to Pax River ***

5507	156513	P-3C	AMARC - FLA	NUD
5508	156514	EP-3E	USN / CILOP	(ARIES II)
5509	156515	P-3C	USN	Update III MOD
5510	156516	P-3C	USN	Update III MOD
5511	156517	EP-3E	USN	(ARIES II)
5512	156518	P-3C	USN	Update III MOD
5513	156519	EP-3E	USN / CILOP	(ARIES II)
5514	156520	P-3C	USN	NUD
5515	156521	P-3C	USN	Update III MOD
5516	156522	P-3C	USN	Update III MOD
5517	156523	P-3C	USN	Update III MOD
5518	156524	P-3C	USN	NUD
5519	156525	P-3C	USN	NUD
5520	156526	P-3C	AMARC - FLA	NUD
5521	156527	P-3C	USN	Update III MOD
5522	156528	EP-3E	USN / CILOP	(ARIES II)
5523	156529	EP-3E	USN / CILOP	(ARIES II)
5524	156530	P-3C	USN	Update III MOD
5301	156599	P-3B	SPAIN	(#22-32 / P.3-9)

*** produced as a HW Bravo for Norwegian Air Force (one of five) the aircraft was later sold by the RNoAF to the Spanish Air Force ***

A US Customs P-3 AEW&C (#155299) *Lockheed*

The ORACL Orion (#155300) *RAAF/ DSTO*

A fleet P-3C Update III (#156510) ; seen here in old VP-45 markings. *B. Rogers*

An EP-3E ARIES II (#156517) ; seen here as a testbed aircraft. ***Picciani via Isham collection***

A P-3C Update III (#156523) ; seen here in old VP-40 markings. ***D. Brown***

A Spanish P-3B (#156600) ; seen here in early Spanish color scheme and markings. ***Marco Borst***

A P-3C NUD (#156520) ; seen here with VP-56. ***Lockheed***

A Spanish P-3B (#156602) ; seen here as a Norwegian Bravo. ***Lockheed***

5302	156600 P-3B	SPAIN	(#22-33 / P.3-10)
5303	156601 P-3B	SPAIN	(#22-34 / P.3-11)
5304	156602 P-3B	SPAIN	(#22-35 / P.3-12)

*** as a Norwegian Bravo, the aircraft had a mid-air collision with a Soviet interceptor fighter on 10/13/87 with slight damage to the fuselage and an engine prop ***

5305	156603 P-3N	NORWAY	(#6603)

*** produced as a HW Bravo for Norway, the aircraft was later modified by NADEP Jax into a training / logistics aircraft ***

5525	157310 P-3C	USN	Update III MOD
5526	157311 P-3C	USN	Update III MOD
5527	157312 P-3C	USN	Update III MOD
5528	157313 P-3C	USN	Update III MOD
5529	157314 P-3C	USN	Update III MOD
5530	157315 P-3C	USN	Update III MOD
5531	157316 EP-3E	USN	(ARIES II)
5532	157317 P-3C	USN	Update III MOD
5533	157318 EP-3E	USN	(ARIES II)
5534	157319 EP-3E	USN	Update III MOD
5535	157320 EP-3E	USN	(ARIES II)
5536	157321 P-3C	USN	Update III MOD
5537	157322 P-3C	USN	Update III MOD
5538	157323 P-3C	USN	Update III MOD

A P-3C Update III MOD (#157314) ***Baldur Sveinsson***

A P-3C Update III MOD (#157320) ; seen here as a NUD with VP-31. ***Lockheed***

A P-3C Update III MOD (#157330) ; severely damaged in crash, it was refurbished with Bravo Orion wings. ***US Navy***

5539	157324	P-3C	USN	Update III MOD
5540	157325	EP-3E	USN	(ARIES II)
5541	157326	EP-3E	USN	(ARIES II)
5542	157327	P-3C	USN	Update III MOD
5543	157328	P-3C	USN	Update III MOD
5544	157329	P-3C	USN	Update III MOD

*** produced as a P-3C, the aircraft was later upgraded to the update III configuration- then equipped with the OASIS over-the-horizon targeting system ***

5545	157330	P-3C	USN	Update III MOD

*** produced as a P-3C and later upgraded to the update III configuration - the aircraft crash Landed (wheels up) at NAS Cecil Field on 2/6/88 after the aircraft's landing gear was sheared off after an aborted landing on a runway under construction (crash damaged included the underbelly of the fuselage and both wings) the aircraft was repaired by NADEP Jacksonville which included wings donated by Australia P-3B A9-300 (#155300) and nicknamed "Phoenix - Pride of NADEP" and completed in November 1990 ***

5546	157331	P-3C	USN	Update III MOD
5547	157332	P-3C	STRIKE	

*** mid-air collision with a NASA Convair 990 over Moffett Field with VP-47 on 4/12/73 ***

5548	158204	NP-3C	USN	

*** produced as a P-3C, the aircraft was later acquired by the Naval Air Development Center in Warminster Pa. and modified into a non-acoustic testbed Orion with 2 optical windows installed in the aft section floor used to house a laser generator / receiver sensor as part of the "Tactical Airborne Laser Communications program" - more recently used to test the "Digital MAD" system ***

5549	158205	P-3C	USN	Update III MOD
5550	158206	P-3C	USN	NUD / SMILS

*** produced as a P-3C, the aircraft was later acquired by VX-1 and equipped with the "SMILS" sonobuoy missile impact location system used for east coast missile tests ***

5552	158207	P-3C	USN	Update III MOD
5553	158208	P-3C	USN	Update III MOD
5554	158209	P-3C	USN	Update III MOD
5555	158210	P-3C	USN	Update III MOD
5556	158211	P-3C	USN	Update III MOD
5557	158212	P-3C	USN	Update III MOD
5558	158213	P-3C	STRIKE	

*** collision with tram wires in Pago Pago, with VP-50 on 4/17/80 ***

A P-3C NUD (#158206) ; utilized by VX-1 and equipped with SMILS. ***S. Miller via Isham collection***

5559	158214 P-3C	USN	Update III MOD
5560	158215 P-3C	USN	Update III MOD
5561	158216 P-3C	USN	Update III MOD
5562	158217 P-3C	STRIKE	

*** engine failure causing ditch off Oman with VP-47 on 3/26/95 ***

5563	158218 P-3C	USN	Update III MOD
5564	158219 P-3C	USN	Update III MOD
5565	158220 P-3C	USN	Update III MOD
5566	158221 P-3C	USN	Update III MOD
5567	158222 P-3C	USN	Update III MOD
5568	158223 P-3C	USN	Update III MOD
5569	158224 P-3C	USN	Update III MOD
5570	158225 P-3C	USN	Update III MOD
5571	158226 P-3C	USN	Update III MOD
5551	158227 NP-3D	NRL	(Magnet)

*** produced as the only production RP-3D built by Lockheed for the US Navy's VXN-8 where it was nicknamed "Roadrunner" and also "Pisano Tres" and utilized for Magnetic Survey projects until VXN-8 retired and the aircraft was acquired by NRL- recently re-designated a NP-3D ***

5572	158563 P-3C	USN	Update III MOD
5573	158564 P-3C	USN	Update III MOD
5574	158565 P-3C	USN	Update III MOD
5575	158566 P-3C	USN	Update III MOD
5576	158567 P-3C	USN	Update III MOD
5577	158568 P-3C	USN	Update III MOD
5578	158569 P-3C	USN	Update III MOD
5579	158570 P-3C	USN	Update III MOD
5580	158571 P-3C	USN	Update III MOD
5581	158572 P-3C	USN	Update III MOD
5582	158573 P-3C	USN	Update III MOD
5583	158574 P-3C	USN	Update III MOD
5584	158912 P-3C	USN	

*** produced as a fleet P-3C, the aircraft was later brought up to the Update III configuration and utilized by the force warfare air test directorate "FWATD" (now NAWC Pax River) as an air test range testbed aircraft ***

A P-3C Update III MOD (#158569) ***Baldur Sveinsson***

A P-3C Update III MOD (#158912) ; utilized by NAWC Pax River as a range test aircraft. ***US Navy***

A P-3C Update III MOD (#158919) ; seen here in newer VP-5 markings. ***US Navy***

A P-3C Update III MOD (#159319) ***Baldur Sveinsson***

5585	158913 P-3C	USN	Update III MOD
5586	158914 P-3C	USN	Update III MOD
5587	158915 P-3C	USN	Update III MOD
5588	158916 P-3C	USN	Update III MOD
5589	158917 P-3C	USN	Update III MOD
5590	158918 P-3C	USN	Update III MOD
5591	158919 P-3C	USN	Update III MOD
5592	158920 P-3C	USN	Update III MOD
5593	158921 P-3C	USN	Update III MOD
5594	158922 P-3C	USN	Update III MOD
5595	158923 P-3C	USN	Update III MOD
5596	158924 P-3C	USN	Update III MOD
5597	158925 P-3C	USN	Update III MOD
5598	158926 P-3C	USN	Update III MOD
5599	158927 P-3C	USN	Update III MOD

5600 158928 P-3C USN Update I

*** produced as the P-3C Update I prototype / development aircraft with OMEGA software ***

5601 158929 P-3C E-Systems SRP Prototype

*** produced as a fleet P-3C, the aircraft was later brought up to the Update III configuration - was recently selected as the SRP Prototype aircraft from E-Systems ***

5602 158930 P-3C STRIKE

*** mid-air collision off San Diego with another squadron P-3, VP-50 on 3/21/91***

5603	158931 P-3C	USN	Update III MOD
5604	158932 P-3C	USN	Update III MOD
5605	158933 P-3C	USN	Update III MOD
5606	158934 P-3C	USN	Update III MOD
5607	158935 P-3C	USN	Update III MOD
5608	159318 P-3C	USN	Update III MOD
5609	159319 P-3C	USN	Update III MOD

5610	159320 P-3C	USN	Update III MOD
5611	159321 P-3C	USN	Update III MOD
5612	159322 P-3C	USN	Update III MOD
5613	159323 P-3C	USN	Update III MOD
5614	159324 P-3C	USN	Update III MOD
5615	159325 P-3C	STRIKE	

*** mid-air collision off San Diego with another squadron P-3, VP-50 on 3/21/91 ***

5616	159326 P-3C	USN	Update III MOD
5617	159327 P-3C	USN	Update III MOD
5618	159328 P-3C	USN	Update III MOD
5619	159329 P-3C	USN	Update III MOD
6001	159342 P-3F	IRAN	(#5-8701)

*** produced by Lockheed for the Imperial Iranian Air Force, prior to the Islamic Revolution, the aircraft (six in all) are based on the P-3C but were equipped avionics and systems of the P-3 Bravo - since the revolution, lack of logistical support has grounded several aircraft becoming parts lockers for the remaining flying elements of which were seen during the Gulf War with Iraq ***

6002	159343 P-3F	STRIKE	(#5-8702)

*** presumed lost during the Iran-Iraq War - this aircraft was known to have the HARPOON missile system installed on board ***

6003	159344 P-3F	IRAN	(#5-8703)
6004	159345 P-3F	IRAN	(# 5-8704)
6005	159346 P-3F	IRAN	(#5-8705)
6006	159347 P-3F	IRAN	(#5-8706)
5620	159503 P-3C	USN	Update III MOD
5621	159504 P-3C	USN	Update I
5623	159505 P-3C	USN	Update I
5624	159506 P-3C	USN	Update I
5625	159507 P-3C	USN	(Outlaw Hunter)

*** produced as a P-3C Update I the aircraft was selected as the Update III OTH-T prototype known as Outlaw Hunter that was successfully operational tested during the Gulf War - the aircraft has currently been upgraded to the OASIS III configuration ***

5626	159508 P-3C	USN	Update I
5627	159509 P-3C	USN	Update I
5628	159510 P-3C	USN	Update I

A P-3F Orion (#159347) *Lockheed*

A P-3C Update I (#159508) ; seen here in VP-19 markings. *M. Wada via Baldur Sveinsson*

A WP-3D with NOAA (#159875) *NOAA*

5629	159511	P-3C	AMARC - FLA	Update I
5630	159512	P-3C	USN	Update I
5631	159513	P-3C	USN	Update I
5632	159514	P-3C	USN	Update I
5622	159773	WP-3D	NOAA	(#N42RF)

*** produced by Lockheed for the US Commerce Dept./NOAA as (one of two) production weather reconnaissance and research aircraft capable of conducting both meteorological and oceanographic research - the aircraft supports the seasonal hurricane research effort flying into hurricanes to collect vital research data for the National Hurricane Center ***

5633	159875	WP-3D	NOAA	(#N43RF)
5634	159883	P-3C	AMARC - FLA	Update I
5635	159884	P-3C	USN	Update I
5636	159885	P-3C	USN	Update III MOD
5637	159886	P-3C	USN	Update I
5638	159887	P-3C	USN	Update III MOD
5639	159888	P-3C	AMARC - FLA	Update I
5640	159889	P-3C	USN	

*** produced as a fleet P-3C Update I the aircraft was later utilized as the Update II prototype and later the first Update III demonstration prototype aircraft ***

5641	159890	P-3C	AMARC - FLA	Update I
5642	159891	P-3C	USN	Update III MOD
5643	159892	P-3C	STRIKE	

*** ditched due to overspeed engine-fire off Adak Alaska with VP-9 on 10/26/78 ***

5644	159893	P-3C	USN	Update I
5645	159894	P-3C	USN	Update III MOD
5646	160283	P-3C	USN	Update III MOD
5647	160284	P-3C	USN	Update I
5648	160285	P-3C	USN	Update I
5649	160286	P-3C	USN	Update III MOD
5650	160287	P-3C	USN	Update III MOD
5651	160288	P-3C	USN	Update I
5652	160289	P-3C	USN	Update I

A P-3C stored with others at AMARC (#159888) ; seen here with VP-9 markings. ***W. Whited via Isham collection***

A P-3C Update I/II (#159889) ; seen here with NADC. ***Lockheed***

A P-3C Update III MOD (#160283) ; seen here in VP-46 markings. ***Baldur Sveinsson***

A P-3C Update III MOD (#160287) ; seen here in VP-46 markings. ***Baldur Sveinsson***

A P-3C Update II (#160290) ; seen here with NATC force warfare Air Test Directorate "FWATD" range aircraft. ***D. Springiaid via Isham collection***

A P-3C Update II (#160291) ; seen here with VX-1. ***US Navy***

An Australian P-3C (#160754) ; ditched into the sea off Cocos Island in 1991. ***RAAF***

5653	160290 P-3C	USN	

*** produced as the first production P-3C Update II, the aircraft was later upgraded to the Update II configuration and is now utilized by the force warfare air test directorate (FWATD) as a air test range testbed aircraft ***

5654	160291 P-3C	USN	Update II
5655	160292 P-3C	USN	

*** produced as a fleet P-3C Update II, the aircraft was selected as the Boeing Update IV prototype and modified to that configuration - the Update IV program was later canceled by the Navy and the aircraft is now underwent a special modification for special projects units ***

5656	160293 P-3C	USN	Update III MOD
xxxx	160294 P-3C	xxxxxx	

*** produced initially as an Update II for the US Navy, the aircraft was re-numbered on the production line as 160751 and became the first P-3C built and delivered to Australia ***

5659	160610 P-3C	USN	Update II
5661	160611 P-3C	USN	Update II
5663	160612 P-3C	USN	Update II
5657	160751 P-3C	AUSTRALIA	(# A9-751)

*** was first of ten P-3C Update II delivered to the Royal Australian Air Force - later upgraded to the Australian Update II.5 configuration and modified with the Marconi ASQ-901 acoustic processing and display system ***

5658	160752 P-3C	AUSTRALIA	(# A9-752)
5660	160753 P-3C	AUSTRALIA	(# A9-753)
5662	160754 P-3C	STRIKE	(# A9-754)

*** leading edge failure - ditched off the Cocos Islands with the RAAF on 4/26/91 ***

5664	160755 P-3C	AUSTRALIA	(# A9-755)
5666	160756 P-3C	AUSTRALIA	(# A9-756)
5668	160757 P-3C	AUSTRALIA	(#A9-757)
5672	160758 P-3C	AUSTRALIA	(#A9-758)
5674	160759 P-3C	AUSTRALIA	(#A9-759)
5676	160760 P-3C	AUSTRALIA	(#A9-760)
5665	160761 P-3C	USN	Update II
5667	160762 P-3C	USN	Update II
5669	160763 P-3C	USN	Update II
5671	160764 P-3C	USN	Update II

A P-3C Update II (#161001) ; seen here with VP-MAU over the Maine seashore. ***US Navy***

A P-3C Update II (#161011) ; seen on the frozen flight line in Keflavik, Iceland with VP-26. ***Baldur Sveinsson***

A P-3C Update II (#161123) ; seen here in VP-10 markings. ***B. Rogers via Isham collection***

5673	160765	P-3C	USN	Update II
5675	160766	P-3C	USN	Update II
5670	160767	P-3C	USN	Update II
5677	160768	P-3C	USN	Update II
5678	160769	P-3C	USN	Update II
5679	160770	P-3C	USN	Update II
5680	160999	P-3C	USN	Update II
5681	161000	P-3C	USN	Update II
5683	161001	P-3C	USN	Update II
5684	161002	P-3C	USN	Update II

*** produced as a fleet P-3C Update II, the aircraft was selected as the prototype aircraft for NADEP Jax's phased depot level maintenance program or PDM ***

5685	161003	P-3C	USN	Update II
5686	161004	P-3C	USN	Update II
5687	161005	P-3C	USN	Update II
5688	161006	P-3C	USN	Update II
5690	161007	P-3C	USN	Update II
5691	161008	P-3C	USN	Update II
5692	161009	P-3C	USN	Update II

*** damaged during a sever thunderstorm on 6/13/94, winds spun the aircraft around in the NAS Jacksonville wash rack shearing off the tail section - received a replacement tail empennage from donor P-3B 154585 ***

5694	161010	P-3C	USN	Update II

*** was the 500th P-3 Orion produced by Lockheed and delivered to the USN ***

5695	161011	P-3C	USN	Update II

*** produced as a fleet P-3C Update II, the aircraft was later modified with the OASIS II OTH-T system ***

5696	161012	P-3C	USN	Update II
5698	161013	P-3C	USN	Update II
5699	161014	P-3C	USN	Update II
5700	161121	P-3C	USN	Update II
5701	161122	P-3C	USN	Update II.3

*** produced as a fleet P-3C Update II but with avionics over and above an Update II but missing other systems common to the Update II.5 - in an effort to categorize this configuration the Update II.3 has been used for maintenance purposes ***

5702 161123 P-3C USN Update II

5703 161124 P-3C USN Update II

5705 161125 P-3C USN Update II

5707 161126 P-3C USN Update II

5710 161127 P-3C USN Update II

5713 161128 P-3C USN Update II

5716 161129 P-3C USN Update II

5718 161130 P-3C USN Update II

5721 161131 P-3C USN Update II

5724 161132 P-3C USN Update II.5

*** produced as the first production P-3C Update II.5 by Lockheed ***

7001 161267 P-3C JAPAN (# 5001)

*** produced as P-3C Update II.5 for Japan as part of a larger contract for licensing production of P-3 in Japan by Kawasaki Heavy Industries ***

7002 161268 P-3C JAPAN (# 5002)

7003 161269 P-3C JAPAN (# 5003)

5726 161329 P-3C USN Update II.5

5727 161330 P-3C USN Update II.5

5728 161331 P-3C USN Update II.5

5729 161332 P-3C USN Update II.5

5730 161333 P-3C USN Update II.5

5731 161334 P-3C USN Update II.5

5732 161335 P-3C USN Update II.5

5734 161336 P-3C USN Update II.5

5735 161337 P-3C USN Update II.5

5736 161338 P-3C USN Update II.5

5738 161339 P-3C USN Update II.5

5739 161340 P-3C USN Update II.5

5733 161368 P-3C NETHERLANDS (# 300)

*** produced as P-3C Update II.5 (the first of thirteen) for the Royal Netherlands Navy ***

5737 161369 P-3C NETHERLANDS (# 301)

5741 161370 P-3C NETHERLANDS (# 302)

5745 161371 P-3C NETHERLANDS (# 303)

A P-3C Update II (#161128); seen here with other P-3C Update II and II.5 Orions representing Commander Patrol Wings Five, NAS Brunswick, Maine. *US Navy*

A P-3C Update II.5 (#161335) ; seen here in VP-92 markings. *Baldur Sveinsson*

A Netherlands P-3C Update II.5 (#161372) ; seen here with another Dutch Orion. *Baldur Sveinsson*

A Netherlands P-3C Update II.5 (#161379) ; seen here in older Dutch markings. ***Marco Borst***

A P-3C Update III (#161762) destroyed in a crash in 1991; seen here in VX-1 markings. ***D. Brown***

A P-3C Update III (#161766) ; seen here with a reserve P-3A Orion. ***Lockheed***

5750	161372	P-3C	NETHERLANDS (# 304)	
5754	161373	P-3C	NETHERLANDS (# 305)	
5758	161374	P-3C	NETHERLANDS (# 306)	
5762	161375	P-3C	NETHERLANDS (# 307)	
5765	161376	P-3C	NETHERLANDS (# 308)	
5769	161377	P-3C	NETHERLANDS (# 309)	
5773	161378	P-3C	NETHERLANDS (# 310)	
5774	161379	P-3C	NETHERLANDS (# 311)	
5776	161380	P-3C	NETHERLANDS (# 312)	
5740	161404	P-3C	USN	Update II.5
5742	161405	P-3C	USN	Update II.5
5743	161406	P-3C	USN	Update II.5
5744	161407	P-3C	USN	Update II.5
5746	161408	P-3C	USN	Update II.5
5747	161409	P-3C	USN	Update II.5
5748	161410	P-3C	USN	(Special Purpose)

*** produced as a fleet P-3C Update II.5, the aircraft was later upgraded as the second Update III prototype - the aircraft is now been specially modified with unique elongated ventral canoe pod for use by NAWC Headquarter's special projects ***

5749	161411	P-3C	USN	Update II.5
5751	161412	P-3C	USN	Update II.5

5752	161413 P-3C	USN	Update II.5
5753	161414 P-3C	USN	Update II.5
5755	161415 P-3C	USN	Update II.5

*** was utilized as the ISAR radar testbed aircraft during suitability evaluations ***

5756	161585 P-3C	USN	Update II.5
5757	161586 P-3C	USN	Update II.5
5759	161587 P-3C	USN	Update II.5
5760	161588 P-3C	USN	Update II.5
5761	161589 P-3C	USN	Update II.5
5763	161590 P-3C	USN	Update II.5
5764	161591 P-3C	USN	Update II.5
5766	161592 P-3C	USN	Update II.5
5767	161593 P-3C	USN	Update II.5
5768	161594 P-3C	USN	Update II.5
5770	161595 P-3C	USN	Update II.5
5771	161596 P-3C	USN	Update II.5
5772	161762 P-3C	STRIKE	

*** hard landing - destroyed during touch & go's at Crows Field with VP-31 on 9/25/90 ***

*** produced as the first production P-3C Update III aircraft by Lockheed ***

5775	161763 P-3C	USN	Update III
5777	161764 P-3C	USN	Update III
5779	161765 P-3C	USN	Update III
5781	161766 P-3C	USN	Update III
5783	161767 P-3C	USN	Update III
5786	162314 P-3C	USN	Update III
5788	162315 P-3C	USN	Update III
5790	162316 P-3C	USN	Update III
5793	162317 P-3C	USN	Update III
5794	162318 P-3C	USN	Update III
5778	162656 P-3C	AUSTRALIA	(# A9-656)

*** produced as a P-3C Update II.5 (one of ten) for the Royal Australian Air Force ***

A P-3C Update III (#162773) ; seen here in new VP-66 markings.
Baldur Sveinsson

5780	162657 P-3C	AUSTRALIA	(# A9-657)
5782	162658 P-3C	AUSTRALIA	(# A9-658)
5784	162659 P-3C	AUSTRALIA	(# A9-659)
5785	162660 P-3C	AUSTRALIA	(# A9-660)
5787	162661 P-3C	AUSTRALIA	(# A9-661)

*** was utilized as the T56 "smoke reduction" engine mod testbed aircraft ***

5789	162662 P-3C	AUSTRALIA	(# A9-662)
5791	162663 P-3C	AUSTRALIA	(# A9-663)
5793	162664 P-3C	AUSTRALIA	(# A9-664)
5795	162665 P-3C	AUSTRALIA	(# A9-665)

*** was utilized as an experimental paint coating / paint scheme testbed aircraft ***

5796	162770 P-3C	USN	

*** produced as a fleet P-3C Update III and later used as the GPS / CP-2044 prototype aircraft - the aircraft is now utilized by the force warfare air test directorate (FWATD) as an air test range testbed aircraft ***

5797	162771 P-3C	USN	Update III
5798	162772 P-3C	USN	Update III
5799	162773 P-3C	USN	Update III
5800	162774 P-3C	USN	Update III
5801	162775 P-3C	USN	Update III
5802	162776 P-3C	USN	Update III
5803	162777 P-3C	USN	Update III
5804	162778 P-3C	USN	AIP Testbed

A P-3C Update III (#163005) ; seen here with VP-62. ***Terry Taylor***

A Norwegian P-3C Update III (#163296) ***Marco Borst***

*** produced as a fleet P-3C Update III, the aircraft is now utilized by the force warfare air test directorate (FWATD) as an air test range testbed aircraft - this airplane is scheduled to be utilized as the AIP program testbed aircraft ***

5805	162998 P-3C	USN	Update III
5806	162999 P-3C	USN	Update III
5807	163000 P-3C	USN	Update III
5808	163001 P-3C	USN	Update III
5809	163002 P-3C	USN	Update III
5810	163003 P-3C	USN	Update III
5811	163004 P-3C	USN	Update III
5812	163005 P-3C	USN	Update III
5813	163006 P-3C	USN	Update III
5814	163289 P-3C	USN	Update III
5815	163290 P-3C	USN	Update III
5816	163291 P-3C	USN	Update III
5821	163292 P-3C	USN	Update III
5822	163293 P-3C	USN	Update III
5823	163294 P-3C	USN	Update III
5824	163295 P-3C	USN	Update III

Korean P-3C Update III, first of the Lockheed Georgia production line (#165098) ***Lockheed***

Canadian CP-140 Aurora (#140101) ; prototype aircraft seen here with eight of her sister-ships ***CF /DND via VPI***

*** produced as a fleet P-3C Update III, this aircraft was the last P-3 built for the US Navy by Lockheed in its California based factories ***

5817	163296	P-3C	NORWAY	(# 3296)

*** produced as a P-3C Update III (one of four) for the Royal Norwegian Air Force ***

5818	163297	P-3C	NORWAY	(# 3297)
5819	163298	P-3C	NORWAY	(# 3298)
5820	163299	P-3C	NORWAY	(# 3299)
5825	164467	P-3C	PAKISTAN	

*** produced as a P-3C Update II.75 for PAKISTAN (# 25), one of three aircraft now in storage at AMARC until resolution of the Congressionally imposed "Pressler Sanctions" - although configured like a US Navy Update III, the aircraft have different radar system and communications equipment / antenna arrangement ***

5826	164468	P-3C	PAKISTAN	

*** produced as a P-3C Update II.74 for PAKISTAN (# 26) ***

5827	164469	P-3C	PAKISTAN	

*** produced as a P-3C Update II.75 for PAKISTAN (# 27) ***

5831	165098	P-3C	KOREA	(# 950901)

*** produced as a P-3C Update III +for Korea, the first of eight built by Lockheed at its Marietta, Georgia facility - although configured like a US Navy Update III the aircraft has a different radar unit and the airframe is built with more modern manufacturing techniques ***

5832	165099	P-3C	KOREA	(# 950902)
5833	165100	P-3C	KOREA	(# 950903)
5834	165101	P-3C	KOREA	(# 950905)
5835	165102	P-3C	KOREA	(# 950906)
5836	165103	P-3C	KOREA	(# 950907)
5837	165104	P-3C	KOREA	(# 950908)
5838	165105	P-3C	KOREA	(# 950909)

CP-140 AURORA

5682	140101	CP-140	CANADA	(# CP-101)

*** produced as the first of eighteen CP-140 AURORA, the aircraft was utilized as a Lockheed air test vehicle developing the Aurora's unique tactical suite - the CP-140 is based on a P-3C with systems, sensors and avionics of the Lockheed S-3A Viking ***

5689	140102	CP-140	CANADA	(# CP-102)
5693	140103	CP-140	CANADA	(# CP-103)
5697	140104	CP-140	CANADA	(# CP-104)
5704	140105	CP-140	CANADA	(# CP-105)

Canadian CP-140A Arcturus (#140119); seen here on the ramp with Arcturus #121. ***P-3 Publications***

5706	140106	CP-140	CANADA	(# CP-106)
5708	140107	CP-140	CANADA	(# CP-107)
5709	140108	CP-140	CANADA	(# CP-108)
5711	140109	CP-140	CANADA	(# CP-109)
5712	140110	CP-140	CANADA	(# CP-110)
5714	140111	CP-140	CANADA	(# CP-111)
5715	140112	CP-140	CANADA	(# CP-112)
5717	140113	CP-140	CANADA	(# CP-113)
5719	140114	CP-140	CANADA	(# CP-114)
5720	140115	CP-140	CANADA	(# CP-115)
5722	140116	CP-140	CANADA	(# CP-116)
5723	140117	CP-140	CANADA	(# CP-117)
5725	140118	CP-140	CANADA	(# CP-118)

CP-140A ARCTURUS

5828	140119	CP-140	CANADA	(# CP-119)

*** produced as a variant of the CP-140 Aurora, the CP-140A ARCTURUS is a specialized reconnaissance aircraft utilized for sovereignty patrols of Canada's northern territories as well as other missions including ice reconnaissance, pollution monitoring, animal population census, Arctic SAR and pilot training - the aircraft differs from the Aurora in the lack of ASW equipment and the installation of specialized radar and enhanced long range navigation / communication gear ***

5829	140120	CP-140	CANADA	(# CP-120)
5830	140121	CP-140	CANADA	(# CP-121)

KAWASAKI P-3 ORION PRODUCTION

SERIAL No.	Type	Location	Comments
5004 - 5008	P-3C	JAPAN	Update II.5(Knockdowns)

*** assembled from Lockheed built "knockdown" components by Kawasaki Heavy Industries ***

SERIAL No.	Type	Location	Comments
5009	P-3C	JAPAN	Update II.5

*** produced as the first Kawasaki full production P-3 Orion under license from Lockheed ***

SERIAL No.	Type	Location	Comments
5010 - 5069	P-3C	JAPAN	Update II.5
(5032)	P-3C		STRIKE

*** wheels up landing at Iwo Jima with the JMSDF on 3/21/92 ***

SERIAL No.	Type	Location	Comments
5070	P-3C	JAPAN	Update III

*** produced as the first Kawasaki P-3C Update III production aircraft ***

SERIAL No.	Type	Location	Comments
5071- 5099	P-3C	JAPAN	Update III
5100	P-3C	JAPAN	Update III+(Proposed Production)

*** planned production of a P-3C Update III+, that adds the proposed CP-2044 computer (under license), an imaging radar and domestically produced GPS and SATCOM avionics ***

SERIAL No.	Type	Location	Comments
xxxx	P-3C	JAPAN	Update III+(Proposed)

*** planned production of a P-3C Update III+, to replace aircraft lost in 1992 ***

SERIAL No.	Type	Location	Comments
9171	EP-3	JAPAN	(ELINT Variant prototype)

*** produced as a electronic signals intelligence collection aircraft similar to the US Navy's
EP-3E Orion ***

SERIAL No.	Type	Location	Comments
9151U	P-3C	JAPAN	(In-flight Testbed Aircraft)

*** produced as a utility version of the Kawasaki P-3C, the aircraft is utilized as an avionics and sensor systems testbed ***

SERIAL No.	Type	Location	Comments
xxxx	EP-3	JAPAN	(Proposed Production)
xxxx - xxxx	UP-3D	JAPAN	(Proposed Production)

*** planned production of another Kawasaki P-3C variant designated UP-3D, the aircraft (two to be built) are to be utilized in electronic warfare training to Japanese surface fleet elements - the mission is similar to the American EP-3J electronic counter measures/counter-counter measures fleet training Orions ***

SERIAL No.	Type	Location	Comments
xxxx	EP-3	JAPAN	(Proposed Production)

P-3 ORION PRODUCTION

US NAVY ORIONS :		
	P-3A	158
	P-3B	125
	P-3C	266
	RP-3D	1
	Total Aircraft :	**550**
FOREIGN ORIONS :		
New Zealand	P-3B	6 (Plus one P-3B from Australian stocks)
Australia	P-3B	10(one to NZ and six to Portugal - one strike)
	P-3C	20(one strike)
	TAP-3	3 (all from USN stocks)
Norway	P-3B	7 (two from USN stocks - five sold to Spain)
	P-3C	4
Netherlands	P-3C	13
Canada	CP-140	18
	CP-140A	3
Japan	P-3C	3
Iran	P-3F	6
Spain	P-3A	7 (all from USN stock)
	P-3B	5 (all from Norway)
Portugal	P-3P	6 (all from Australian stock via Lockheed)
Chile	UP-3A	8 (all from USN stock)
Thailand	P-3T	2 (both from USN stock)
	UP-3T	1 (from USN stock)
Greece	P-3B	6 (from USN stock)
Pakistan	P-3C	3
Commerce Dept.	WP-3D	2
	Total Aircraft :	**642**
Korea	P-3C	8 (current Lockheed production)
Kawasaki H/I (Japan)	P-3C	101
	EP-3	3 (two more aircraft proposed)
	UP-3C	1
	UP-3D	- (two aircraft proposed)
	UP-3E	- (several aircraft planned from JMSDF stock)
	Total Aircraft :	**755**

US Navy Patrol Squadrons

SQD	Nickname	Command Wing	Base - Station	Tail Code
VP-1	Screaming Eagles	COMPATWINGTENNAS	Whidbey Island, WA.	(YB)
VP-4	Skinny Dragons	COMPATWINGTWONAS	Barbers Point, HI.	(YD)
VP-5	Mad Fox	COMPATWINGELEVEN	NAS Jacksonville, FL	(LA)
VP-6	Blue Sharks *	COMPATWINGTWO	NAS Barbers Point, HI	(PC)
VP-8	Tigers	COMPATWINGFIVE	NAS Brunswick, ME.	(LC)
VP-9	Golden Eagles	COMPATWINGTWO	NAS Barbers Point, HI	(PD)
VP-10	Red Lancers	COMPATWINGFIVE	NAS Brunswick, ME.	(LD)
VP-11	Pegasusp	COMPATWINGFIVE	NAS Brunswick, ME.	(LE)
VP-16	War Eagles	COMPATWINGELEVEN	NAS Jacksonville, FL.	(LF)
VP-17	White Lightings *	COMPATWINGTWO	NAS Barbers Point, HI	(ZE)
VP-19	Big Red *	COMPATWINGTEN	NAS Moffett Field, CA.	(PE)
VP-22	Blue Gueese *	COMPATWINGTWO	NAS Barbers Point, HI.	(QA)
VP-23	Sea Hawks *	COMPATWINGFIVE	NAS Baunswick, ME.	(LJ)
VP-24	Batmen *	COMPATWINGELEVEN	NAS Jacksonville, FL.	(LR)
VP-26	Tridents	COMPATWINGFIVE	NAS Brunswick, ME.	(LK)
VP-28	*	COMPATWINGTWO	NAS Barbers Point, HI.	(QC)
VP-30	Pros Nest	COMPATWINGELEVEN	NAS Jacksonville, FL.	(LL)
VP-31	Black Lightings *	COMPATWINGTEN	NAS Moffett Field, CA.	(RP)
VP-40	Fighting Marlins	COMPATWINGTEN	NAS Whidbey Island, WA.	(QE)
VP-44	Golden Pelicans *	COMPATWINGFIVE	NAS Brunswick, ME.	(LM)
VP-45	Pelicans	COMPATWINGELEVEN	NAS Jacksonville, FL.	(LN)
VP-46	Gray Knights	COMPATWINGTEN	NAS Whidbey Island, WA.	(RC)
VP-47	Golden Swordsmen	COMPATWINGTWO	NAS Barbers Point, HI	(RD)
VP-48	Boomers *	COMPATWINGTEN	NAS Moffett Field, CA.	(SF)
VP-49	Woodpeckers *	COMPATWINGELEVEN	NAS Jacksonville, FL	(LP)
VP-50	Blue Dragons *	COMPATWINGTEN	NAS Moffett Field, CA.	(SG)
VP-56	Dragons	COMPATWINGELEVEN	NAS Jacksonville, FL.	(LQ)
VP-60	Cobras *	RESPATWINGPAC	NAS Chicago, IL.	(LS)
VP-62	Broadarrows	RESPATWINGLANT	NAS Jacksonville, FL.	(LT)
VP-64	Condors	RESPATWINGLANT	NAS Willow Grove, PA.	(LU)
VP-65	Tridents	RESPATWINGPAC	NAS Point Mugu, CA.	(PG)
VP-66	Liberty Bells	RESPATWINGLANT	NAS Willow Grove, PA.	(LV)
VP-67	Golden Hawks *	RESPATWINGPAC	NAS Memphis, TN.	(PL)
VP-68	Black Hawks p	RESPATWINGLANT	NAS Washington, DC.	(LW)
VP-69	Totems	RESPATWINGPAC	NAS Whidbey Island, WA.	(PJ)
VP-90	Lions *	RESPATWINGPAC	NAS Chicago, IL.	(LX)
VP-91	Black Cats	RESPATWINGPAC	NAS Moffett Field, CA.	(PM)
VP-92	Minutemen +	RESPATWINGLANT	NAS Brunswick, ME.	(LY)
VP-93	Executioners *	RESPATWINGLANTNAS	Detroit, MI.	(LH)
VP-94	Crawfishers	RESPATWINGPAC	NAS New Orleans, LA.	(LZ)
VP-MAU	Northern Sabers *	COMPATWINGFIVE	NAS Brunswick, ME.	(LB)
VP-MAU	Rolling Thunders *	COMPATWINGTEN	NAS Moffett Field, CA.	(PS)
VX-1	Pioneers	(Air test & development)	NAS Pax River, MD.	(JA)
VQ-1	World Watchers	(signal intelligence)	NAS Whidbey Island,	(PR)
VQ-2	Batmen	(signal intelligence)	NAS Rota, Spain	(JQ)
VQ-11	Bandits	(C^3I / counter measures)	NAS Brunswick, ME.	
VPU-1	Old Buzzards	(special projects)	NAS Brunswick, ME.	
VPU-2	Wizards	(special projects)	NAS Barbers Point, HI	(SP)
VXN-8	World Travelers *	(ocean development unit)	NAS Pax River, MD	(JB)
VAQ-33	Firebirds *	(fleet C3I/CM training unit)	NAS Key west, FL.	(GD)
VW-4	Hurricane Hunters *	(weather reconnaissance unit)	NAS Jacksonville, FL	(MH)
NRL -		Flight Support Detachment	NAS Pax River, MD.	
NAWC-AD	Pax River	(force warfare air test directorate)	NAS Pax River, MD	
NAWC-AD	Warminster	(now integrated with NAWC)	NAS Pax River, MD.	
NAWC-WD	Point Mugu	(pacific missile test center)	NAS Point Mugu, CA.	

KEY CHART
* Disestablished
p pending retirement
+ recently moved from NAS North Weymouth to NAS Brunswick

International P-3 Orion Squadrons

Country	Squadron	Base - Station	Comment
Australia	#10	RAAF Base Edinburgh, NSW Australia	92 Wing
Australia	#11	RAAF Base Edinburgh, NSW Australia	92 Wing
Canada	#404	CFB Greenwood, Nova Scotia Canada	14 Wing (moat)
Canada	#405	CFB Greenwood, Nova Scotia Canada	14 Wing
Canada	#407	CFB Comox, British Columbia Canada	19 Wing
Canada	#415	CFB Greenwood, Nova Scotia Canada	14 Wing
Chile	VP-1	Vina del Mar, Chile	
Greece			
Iran	Tactical Air Base #7, Shiras Iran		
Japan	VP-1	JMSDF Base Kanoya	Wing 1
Japan	VP-2	JMSDF Base Hachinohe	Wing 2
Japan	VP-3	JMSDF Base Atsugi	Wing 4
Japan	VP-4	JMSDF Base Hachinohe	Wing 2
Japan	VP-5	JMSDF Base Naha	Wing 5
Japan	VP-6	JMSDF Base Atsugi	Wing 4
Japan	VP-7	JMSDF Base Kanoya	Wing 1
Japan	VP-8	JMSDF Base Iwakuni	Wing 31
Japan	VP-9	JMSDF Base Naha	Wing 5
Japan	VX-81		
Japan	VX-51	JMSDF Base Atsugi	Wing 4 (ADS)
Japan	VP-206	JMSDF Base Shimofusa	Wing 21(ATS)
Netherlands	VS-320	NAS Valkenburg, The Netherlands	
Netherlands	VS-321	NAS Valkenburg, The Netherlands	
Norway	#333	NAS Andoya, Norway	
New Zealand	#5	NAS Whenuapai, New Zealand	
Pakistan	#29	Base Sharea Faisal	
Portugal	#601		
Spain	#221	SAF Station Moron, Spain	22 Wing
Thailand	#101	RTNAB Utapao, Thailand Air	Wing #61

-key-

ADS - JMSDF's "Air Development Squadron
ATS - JMSDF's "Air Training Squadron

- KEY LIST -

AMARC- Aircraft Maintenance And Rejuvenation Center located at Davis-Monthan AFB Tucson, Arizona - the US military's desert aircraft storage facility. There are several categories of aircraft storage and disposition at AMARC for Navy aircraft and P-3 Orion in particular.

* 4850-1Force Level Assurance (FLA) aircraft kept in ready condition as reserve material supply - used to be known as war reserve

* 4850-2Foreign Military Sales (FMS) aircraft assigned to the Navy's FMS office for potential sales to other foreign military operators

* 4850-3 Museum assigned to the National Museum of Aviation at Pensacola, Florida. these aircraft are managed by the Pensacola museum and are put on display there or provided to other bases for static display (in some cases, this category is used to trade aircraft to and from civilian collectors)

* 4850-4 Reclamation (REC) aircraft scrapped, equipment removed for spare parts and the remaining airframe disposed of - SARDIP includes a sub-category known as RIT or Reclamation Insurance Type which are aircraft kept as "parts birds" to insure a ready supply of needed spares

AERO UNION - an airborne fire fighting company in Chico, California (pioneered the use of P-3 Orions as air tankers)

AIMD- Aircraft Intermediate Maintenance Dept.

AIP- ASUW Improvement Program

AMPS- Airborne Multi-sensor Pod System

ARIES- Airborne Reconnaissance Integrated Electronic System for EP-3E Orion (modified on a P-3A airframe)

ARIES II- a standardized EP-3E configuration with upgraded Aries systems (now based on the P-3C airframe)

ASA- Administration Support Aircraft (VIP transport detachment of VP-30 at NAS Jacksonville

BER- Beyond Economical Repair associated with SRP aircraft that are shown not to fit the SRP criteria during pre-inspection phase

CILOPS- Conversion In-Lieu Of Procurement the upgrade modification program of P-3C airframes into "EP-3E ARIES II" Orions. The CILOPS program was re-structured several times and were built bydifferent organizations over the years. They include .....

* AEROMOD - Lockheed's aircraft modification facility in Greenville, South Carolina where the first EP-3E ARIES II conversions were made before the remaining program was transferred to NADEP Alameda in June 1992.

* NADEP Alameda - the naval aircraft repair depot that was selected next to perform conversion of EP-3E ARIES II Orions - the aircraft were transferred again to NADEP Jax for completion

CONUS- aircraft flown only in the continental United States

DIFAR- Directional Frequency Analysis and Recording system

DMRO- Defense Material Re-utilization Office

DOE- Department Of Energy

EATS - Extended Area Test System

EFIS- Electronic Flight Instrumentation System

ESM- Electronic Support Measures system (formerly known as ECM)

ETD- Executive Transport Department a division of NAS Barbers Pt. Hawaii (to be re-located to Kaneohe bay Marine air base when COMPATWINGS moves there in 1996-97)

FRAMP- Fleet Readiness Aviation Maintenance Personnel (aircraft used for ground maintenance training instruction)

FWATD- Force Warfare Air Test Directorate, a division of NAWC Pax River (formerly the naval air test center or NATC)

GPS- Global Positioning System

GSA- Government Services Administration

H&P- Hawkins and Powers Aviation, an airborne fire fighting company in Graybull, Wyoming

IPADS- Improved Processor And Display System a P-3C Update III type acoustic display system for TACNAVMOD P-3B's

ISAR - Inverse Synthetic Aperture Radar

K-Tech- K-Tech Aviation, an aircraft parts company in Tucson, Arizona

KNOCK-DOWNS- aircraft re-assembled from large component pieces

MAD - Magnetic Anomaly Detector

NARA- Naval Air Reserve Activity

NADC- Naval Air Development Center, Warminster Pennsylvania (now incorporated into NAWC Pax River)

NADEP- Naval Aviation Depot (which performs aircraft rework, overhauls, mods and painting) facilities associated with P-3 Orion are located at NASJacksonville, Florida and the former NAS Alameda, California

NASC - Naval Air Systems Command (also known as NAVAIR)

NATC - Naval Air Test Center located at NAS Pax River, Maryland - now NAWC-AD Pax River

NAWC -AD- Naval Air Warfare Center - Aircraft Division located at NAS Pax River NOAA- National Oceanic and Atmospheric Administration

NRL- Naval Research Laboratory

NAWC-23 - Naval Air Warfare Center - Headquarters

NFATS- Naval Force Aircraft Test Squadron is a newly formed unit to provide flight operation to Pax River based NAWC test range aircraft including NP-3D Orions

NWTS- Naval Weapons Test Squadron is a newly formed unit to provide flight operations to Pt. Mugu based NAWC test range aircraft including NP-3D Orions

NP-3D- a newly designated Orion variant (as of February 1, 1995) that encompasses a number of P-3 utilized by NAWC facilities for RDT&E as well as various test range duties

OASIS- Over-the-horizon Sensor Information System

OGMA- Portuguese re-work facility in Alverca Portugal

OTH-T- Over-The-Horizon Targeting system

OUTLAW HUNTER- OTH-T prototype aircraft testbed

PMTC- Pacific Missile Test Center, Point Mugu, California (now NAWC-AD Point Mugu)

SARDIP- Stricken Aircraft Reclamation and Disposal Program (scrapped aircraft stripped of needed parts and sold off as scrap metal)

SATCOM - Satellite Communications

SDLM- Standard Depot Level Maintenance (performed by NADEPS)

SMILS- Sonobuoy Missile Impact Location System

SRP- Sustained Readiness Program

STRIKE- aircraft stricken off Navy books due to accidents, collisions and crashes

TNM - TACNAVMOD, tactical navigation modification a systems upgrade modification to P-3A and P-3B Orions that brought them up to P-3C processing standards

Appendix E
Orion Systems Configuration Matrix Charts

P-3 AVIONICS CONFIGURATIONS

	AVIONICS FUNCTION	P-3A SN 5001-5157 *17	P-3A/P-3B *18 DIFAR RETROFIT	P-3B *19 (TAC/NAV MOD)	P-3C SN 5501-5619	P-3C UPDATE I SN 5620-5652	P-3C UPDATE II SN 5653-5771	P-3C UPDATE III SN5772 *9	P-3C UPDATE III RETROFIT	P-3C UPDATE IV
DPS & DISPLAYS	GEN. PURPOSE DIGITAL COMPUTER	–	–	–	ASQ-114	ASQ-114	ASQ-114	ASQ-114	ASQ-114 + AYK-14(1)	ASQ-114 – AYK-14(1)
	DATA PROCESSING SYSTEM	ASA-16	ASA-16	ASN-124	AYA-8A DPS L.U.(3)	AYA-8B DPS L.U.(4)	AYA-8B DPS L.U.(4)	AYA-8B DPS L.U.(4)	AYA-8B DPS M.L.U (3), LU(4)	AYA 8B DPS AND LU(3) LU(4)
	TACTICAL DISPLAY	ASA-16	ASA-16	ASA-66B	ASA-70	ASA-70	ASA-70	ASA-70	ASA-70	NEW
	NAVIGATION DISPLAY	PT-396	PT-396	PT-396	IP-915 NAV	IP-919 NAV	IP-919 NAV	IP-919 NAV	IP-919 NAV	NEW
	SENSOR DISPLAY	APA-125A	APA-125A	APA-125A	ASA-70	ASA-70	ASA-70	ASA-70	ASA-70	NEW
	PILOT DISPLAY	GTP-4	GTP-4	ASA-66(1)	ASA-66(1)	ASA-66(2)	ASA-66(2)	ASA-66(2)	ASA-66 (2)	NEW
	INTERVAL COMPUTER	TD-441A	TD-441A	TD-441A	GPDC FUNCTION	GPDC FUNCTION	GPDC FUNCTION	GPDC FUNCTION	GPDC FUNCTION	GPDC FUNCTION
	DMTS	–	–	RD-461	RD-319	RD-319A	RD-319A/ASH-33*6	ASH-33A	ASH-33A	ASH-33A
COMMUNICATIONS	UHF TRANSCEIVER	ARC-52/51A(2)	ARC-51A(2)	ARC-51A(2)	ARC-143(2)	ARC-143B(2)	ARC-143B(2)	ARC-143B(2)	ARC-143B(3)	ARC-166(2)
	VHF (AM)	ARC-101	ARC-101	ARC-101	ARC-101	ARC-101	ARC-101/618M-3A*6	618M-3A	618M-3A	618M-3A
	HF TRANSCEIVER	ARC-94(2), ARR-41	ARC-94(2), ARR-41	ARC-94(2), ARR-41	ARC-161(2)	ARC-161(2)	ARC-161(2)	ARC-161(2)	ARC-161(2)	ARC-161(2)
	HF SECURE VOICE *16	KY-75	KY-75	KY-75	KY-75	KY-75	KY-75	KY-75	KY-75	KY-75
	TELETYPE	AGC-5(V)*16	AGC-9(V)*16	AGC-9(V)	AGC-6	AGC-6	AGC-6	AGC-6	AGC-9	AGC-9
	DATA LINK (LINK 11)	–	–	–	ACQ-5 (MODEL 4)	ACQ-5 (MODEL 4)	ACQ-5 (MODEL 4)	ACQ-5 (MODEL 4)	CV-2969/AYC (MODEL 4)	CV-2969/AYC (MODEL 4)
	INTERCOM	AIC-22V	AIC-22V	AIC-22V	AIC-22V	AIC-22V	AIC-22V	AIC-22V	AIC-22(V)	AIC-22(V)
	IFF/SIF	APX-6, APX-7	APX-72, APX-7	APX-72, APX-7	APX-72, APX-76A	APX-72, APX-76A(V)	APX-72, APX-76A(V)	APX-72, APX-76A(V)	APX-72, APX-76A(V)	APX-72, APX-76A(V)
	IACS	–	–	–	OV-78A*16	OV-78A*16	OV-78A*16	OV-78A	OV-78A	OV-78A
	SATCOM (OTCIXS)	–	–	–	–	–	–	–	MD-1049	MC-1049
NAVIGATION	INERTIAL	ASN-42	ASN-42	LTN-72	LTN-72(2) RETRO	LTN-72(2) RETRO	LTN-72(2)*10	LTN-72(2)	LTN-72(2)	LTN-72(2)
	DOPPLER NAVIGATOR	APN-153	APN-153	APN-153	APN-187	APN-187	APN-187/APN-227*6	APN-227	APN-227	APN-227
	OMEGA/LORAN	APN-70 LORAN	APN-70 LORAN 'C'	ARN-99, APN-70	LTN-211 OMEGA	ARN-99 OMEGA	ARN-99 OMEGA	ARN-99 OMEGA	ARN-99 OMEGA	GPS
	AIRMASS COMPUTER	ASA-47	ASA-47	ASN-124	–	–	–	–	–	–
	TAS COMPUTER	A/A-24G-9	A/A-24G-9	A/A-24G-9	A/A-24G-9	A/A-24G-9	A/A-24G-9	A/A-24G-9	A/A-24G-9	A/A-24G-9
	VOR SET	ARN-87(2)	ARN-87(2)	ARN-87(2)	ARN-87(2)	ARN-87(2)	ARN-87(2)/VIR-31A*6	VIR-31A	VIR-31A	VIR-31A
	TACAN	ARN-52(V)	ARN-52(V)	ARN-52(V)	ARN-84 OR ARN-52(V)	ARN-84	ARN-84/ARN-118*6	ARN-118	ARN-118	ARN-118
	ADF	ARN-83	ARN-83	ARN-83	ARN-83	ARN-83	ARN-83	ARN-83	ARN-83	ARN-83
	UHF DF/OTPI	ARA-25A	ARA-25A	ARA-25A	ARA-50	ARA-50	ARA-50	ARA-50	ARA-50	ARA-50
	MARKER BEACON	ARN-32	ARN-32	ARN-32	ARN-32	ARN-32	ARN-32/VIR-31A*6	VIR-31A	VIR-31A	VIR-31A
	FLIGHT DIRECTOR SYSTEM/AHRS	ASN-37	ASN-50	ASN-50	AJN-15	AJN-15	AJN-15	AJN-15	AJN-15	AJN-15
	ILS/GLIDE SLOPE	*1	*1	*1	51 V-4	51 V-4	51 V-4/VIR-31A*6	VIR-31A	VIR-31A	VIR-31A
	SONO REFERENCE SYSTEM	R-1047 OTPI *8	R-1047 OTPI *8	R-1047 OTPI *8	R-1651 OTPI	R-1651 OTPI	ARS-3 SRS	ARS-3 SRS	ARS-3 SRS	SRS (99 CHANNEL)
	AUTO PILOT/AFCS	PB-20N	PB-20N	PB-20N	ASW-31A *3	ASW-31A	ASW-31A	ASW-31A	ASW-31A	ASW-31A
	RADAR ALTIMETER/WARNING	APN-141/APQ-107	APN-141/APQ-107	APN-141/APQ-107	APN-194/APQ-107	APN-194/APQ-107	APN-194/APQ-107	APN-194/APQ-107	APN-194/APQ-107	APN-194/APQ-107
ACOUSTIC	SONOBUOY SIGNAL PROCESSOR	AQA-5	AQA-7(V) 2	AQA-7(V) 5	AQA-7(V) 1	AQA-7(V) 4	AQA-7(V) 8 *11	SASP UYS-1(V)	SASP UYS-1(V)	SASP UYS-1(V)
	SONOBUOY RECEIVER GROUP	ARR-52A	ARR-72	ARR-72	ARR-72	ARR-72	ARR-72	ARR-78(2)	ARR-78(2)	ARR-78(2)
	ACTIVE SONAR PROCESSING	AQA-1, ASA-20	AQA-7	AQA-7	AQA-7	AQA-7	AQA-7	SASP	SASP	SASP
	TAPE RECORDER	UNH-6	AQH-1 OR 4 (V)1	AQH-1 OR 4(V)1	AQH-4 (V)1	AQH-4 (V)1	AQH-4 (V)2	AQH-4 (V)2 (2)	AQH-4 (V)2 (2)	AQH-4 (V)2 (2)
	BATHYTHERMOGRAPH RECORDER	–	RO-308	RO-308	RO-308	RO-308	RO-308	SASP	SASP	SASP
	HYPERBOLIC FIX UNIT	–	–	–	HYFIX	HYFIX	HYFIX	–	–	–
	TIME CODE GENERATOR	–	TD-900/AS	TD-900/AS	TD-900/AS	TD-900A/AS	TD-900A/AS	TD-900A/AS	TD-900A/AS	TD-900A/AS
	SEA NOISE METERS	–	ID-1872 *2	ID-1872 *2	ID-1872/A	ID-1872/A	ID-1872/A	SASP	SASP	SASP
	GEN. XMTR GP (CASS)	–	ASA-76	ASA-76	ASA-76	ASA-76	ASA-76	SASP	SASP	SASP
NONACOUSTIC	RADAR	APS-80B	APS-80B	APS-80B	APS-115B	APS-115B	APS-115B	APS-115B	APS-137	APS-134(V)4
	ESM	ALD-2B, ULA-2	ALD-2B, ULA-2	ALR-66(V)2 *15	ALQ-78	ALQ-78	ALQ-78	ALQ-78	ALR ( )	ALR ( )
	MAD/AUTO DETECT	ASQ-10A	ASQ-10A	ASQ-10A	ASQ-81 *4	ASQ-81(V)1	ASQ-81(V)1	ASQ-81(V)1	ASQ-81(V)1	ASQ-81(V)1
	MAD COMPENSATOR	CN-191	CN-191	CN-191	ASA-65(V)2 *4	ASA-65(V)2	ASA-65(V)5 *13	ASA-65(V)5	ASA-65(V)5	ASA-65(V)5
	SAD	–	–	–	ASA-64	ASA-64	ASA-64	ASA-64	ASA-64	ASA-64
	IRDS	AAS-36 *5	AAS-36 *5	AAS-36	AAS-36 *5	AAS-36 *5	AAS-36	AAS-36	AAS-36	AAS-36
ARMAMENT, PHOTO, MISC.	MISSILES/CONTROL SYSTEM	BULLPUP	BULLPUP	HARPOON	HARPOON	HARPOON	HARPOON	HARPOON	HARPOON	HARPOON
	SURVEILLANCE CAMERA	KS-89	KS-89	KS-89	KA-74	KA-74	–	–	–	–
	SPECIAL PURPOSE CAMERA	–	–	–	KB-18	KB-18	–	–	–	–
	CRASH LOCATOR	–	ASH-20(V)	ASH-20(V)	ASH-20(V)	ASH-20(V)	ASH-20(V)	ASH-20(V)	ASH-20(V)	ASH-20(V)
	SYNCHRO CONVERTOR	–	–	–	CV-2461 A/A	CV-2461 A/A	CV-2461 A/A	CV-2461 A/A	CV-2461 A/A	CV-2461 A/A
										SELF-DEFENSE THREAT WARNING RADAR CHAFF, FLARES, DECOYS

USN P-3 Orion avionics charts (P-3A/B, P-3C NUD and Updates) *Lockheed*

EQUIPMENT	UPDATE II	UPDATE II.5	UPDATE III
DATA PROCESSING AND DISPLAYS			
Digital Data Computer	AN/ASQ-114	AN/ASQ-114	AN/ASQ-114
Data Analysis and Programming	AN/AYA-8B	AN/AYA-8B	AN/AYA-8CB
Logic Units	LU-1,2,3,4	LU-1,2,3,4	LU 1,2,3,LU 4
Magnetic Tape Transport	RD-319A	AN/ASH-33	AN/ASH-33A
Time Code Generator	TD-900A/AS	TD-900A/AS	TD-900A/AS
Tactical Display (Tacco)	IP-917/ASA-70	IP-917/ASA-70	IP-917/ASA-70
Tactical Display (Pilot)	IP-886A/ASA-66	IP-886A/ASA-66	IP-886A/ASA-66
Data Display Control (Pilot)	C-7444/ASA-66	C-7444/ASA-66	C-7444/ASA-66
Tactical Display (Senso)	IP-886A/ASA-66	IP-886A/ASA-66	-
Data Display Control (Senso)	C-7444/ASA-66	C-7444/ASA-66	-
Sonobuoy Display (Senso)	-	-	AN/USQ-78
Auxiliary Readout Display	IP-919/ASA-70	IP-919/ASA-70	IP-919/ASA-70
IRDS Auxiliary Display	OD-159/A	OD-159/A	OD-159/A
Radar Display	IP-918/ASA-70	IP-918/ASA-70	IP-918/ASA-70
Radar Scan Converter	AN/ASA-69	AN/ASA-69	AN/ASA-69
Acoustic Interface Unit (AIU)	-	-	-
Arm/Ord controls	Various	Various	Various
COMMUNICATIONS			
UHF Transceiver	AN/ARC-143B (2)	AN/ARC-143B (2)	AN/ARC-143B (2)
UHF/VHF Tranceiver	AN/ARN-87/AN/ARC-101	AN/ARC-197	AN/ARC-197
HF Tranceiver	AN/ARC-161 (2)	AN/ARC-161 (2)	AN/ARC-161 (2)
Teletype/Secure TTY	AN/AGC-6/KW-7	AN/AGC-6/KW-7	AN/AGC-6/KW-7
Data Terminal/Data Link	AN/ACQ-5	AN/ACQ-5	AN/ACQ-5
Secure Voice	KY-28	KY-28	KY-28, KY-75
IFF Transponder	AN/APX-72	AN/APX-72	AN/APX-72
Transponder Computer	KIT-1A/TSEC	KIT-1A/TSEC	KIT-1A/TSEC
IACS	OV-78/A	OV-78/A	OV-78/A
ICS	AN/AIC-22(V)	AN/AIC-22(V)	AN/AIC-22(V)
Emergency Transmitter	AN/PRT-5	AN/PRT-5	AN/PRT-5
Crash Locater System	AN/URT-26(V)	AN/URT-26(V)	AN/URT-26(V)
Satellite Communications			
ARMAMENT, MISCELLANEOUS			
Harpoon	AN/AWG-19(V)1	AN/AWG-19(V)1	AN/AWG-19(V)1
Bomb Racks/Adapters	AERO-1A Adapters BRU-12/A, 14/A, 15/A	AERO-1A Adapters BRU-12/A, 14/A, 15/A	AERO-1A Adapters BRU-12/A, 14/A, 15/A
APU	GTCP 95-2	GTCP 95-2	GTCP 95-3
EDC	81 HP (2)	81 HP (2)	81 HP (2)
NAVIGATION			
Inertial Navigation Set	LTN-72	LTN-72 (2)	LTN-72 (2)
Omega	AN/ARN-99(V)1	AN/ARN-99(V)1	AN/ARN-99(V)1
GPS			
TACAN	AN/ARN-84	AN/ARN-118	AN/ARN-118
VOR	AN/ARN-87(V) (2)	AN/ARN-140 (2)	AN/ARN-140 (2)
Radar Altimeter	AN/APN-194(V)1	AN/APN-194(V)1	AN/APN-194(V)1
Doppler Navigation Radar	AN/APN-187	AN/APN-227	AN/APN-227
UHF/DF	AN/ARA-50	AN/ARA-50	AN/ARA-50
LF/DF	AN/ARN-83	AN/ARN-83	AN/ARN-83
Flight Director	AN/AJN-15	AN/AJN-15	AN/AJN-15
ILS/Glide Slope	51V4	AN/ARN-140 (2)	AN/ARN-140 (2)
Marker Beacon	AN/ARN-32	AN/ARN-140 (2)	AN/ARN-140 (2)
Attitude Heading Reference	ID-1540/A	ID-1540/A	ID-1540/A
Autopilot	AN/ASW-31	AN/ASW-31	AN/ASW-31
Encoder Altimeter	AAU-32/A (2)	AAU-32/A (2)	AAU-32/A (2)
Sonobuoy Reference System	AN/ARS-3	AN/ARS-3	AN/ARS-3
Compass System	ML-1	ML-1	ML-1
RAWS	AN/APQ-107 (2)	AN/APQ-107 (2)	AN/APQ-107 (2)
OTPI	R-1651/ARA	R-1651/ARA	-
ACOUSTIC			
Sonobuoy Receiver	AN/ARR-72(V)	AN/ARR-72(V)	-
Acoustic Processor	AN/AQA-7(V)8 and 9	AN/AQA-7(V)8 and 9	AN/UYS-1
CASS Generator Transmitter	AN/ASA-76	AN/ASA-76	T-1234/AN/ASA-76
Tape Recorder	AN/AQH-4(V)2	AN/AQH-4(V)2	AN/AQH-4(V)2
Bathythermographic Group	RO-308/SSQ-36	RO-308/SSQ-36	-
Test Signal Generator	SG-791	SG-791	SG-1156/A
Ambient Sea Noise Meter	ID-1872/A	ID-1872/A	-
NON-ACOUSTIC			
Radar	AN/APS-115B (2)	AN/APS-115B (2)	AN/APS-115B (2)
Radar Recognition/SIF	AN/APX-76	AN/APX-76	AN/APX-76
Video Decoder	AN/APX-76	AN/APX-76	AN/APX-76
ESM	AN/ALQ-78	AN/ALQ-78A	AN/ALQ-78A
MAD	AN/ASQ-81	AN/ASQ-81	AN/ASQ-81
MAD Compensater	AN/ASA-65(V)5	AN/ASA-65(V)5	AN/ASA-65(V)5
MAD Recorder	AN/ASA-64	AN/ASA-64	AN/ASA-64
IRDS	AN/AAS-36	AN/AAS-36	AN/AAS-36
Video Recorder	OA-8962/ASH	OA-8962/ASH	OA-8962/ASH

USN P-3C Orion Update II, II.5 and III avionics chart *Lockheed*

TABLE 3-A FMS P-3 AVIONICS CONFIGURATIONS (Continued)

AVIONICS FUNCTION	SAF P-3A	RNZAF P-3B	RNOAF P-3B	RAAF P-3B	RAAF P-3C	RAAF P-3C	RNLN P-3C	JMSDF P-3C	IRAN P-3F	CAF CP-140
ILS/GLIDE SCOPE	51V-4	---	51V-4	---	51V-4	ARN-140	ARN-140	ARN-140	---	ARN-508
OTPI	R-1047	R-1047	R-1047	R-1047	R-1651	HAZELTINE	R-1651	R-1651	(1)	---
SONO REFERENCE	---	---	---	---	ARS-3	ARS-3	ARS-3	ARS-3	---	ARS-501
AUTO PILOT/AFCS	PB-20N	PB-20N	PB-20N	PB-20N	ASW-31	ASW-31A	ASW-31A	ASW-31A	ASW-31	ASW-502
RADAR ALTIMETER	APN-141	APN-141	APN-141	APN-141	APN-194	APN-194	APN-194	APN-194	APN-194	APN-511
RAWS	APQ-107	APQ-107	APQ-107	APQ-107	APQ-107	APQ-107	APQ-107	APQ-107	APQ-107	APN-511
CENTRAL REPEATER	---	---	---	---	AM/4923/A	AM/4923/A	AM/4923/A	AM/4923/A	---	AM-4923/A
ACOUSTIC										
SONO SIG PROCESS	AQA-5	AQA-5	AQA-7(V)2	AQA-5D	AQS-901	AQS-901	AQA-7(V)8/9	AQA-7(V)4/5	(1) AQA-5	OL-5004/AYS
SONOBUOY REC GRP	ARR-52	ARR-52	ARR-72	ARR-52	ARR-901	ARR-901	ARR-72	ARR-72	ARR-75	ARR-76
ACT SONAR PROCESS	AQA-1/ASA-20	AQA-1/ASA-20	AQA-7	AQA-1	AQS-901	AQS-901	(1) UYQ-8	(1) UYQ-8	AQA-1	---
			---	ASA-20	---		---	---		---
TAPE RECORDER	UNH-6	AQH-1	AQH-4(V)2	AQH-1	AQH-4(V)2	AQH-4(V)2	AQH-4(V)2	AQA-4(V)2	AQH-1	AOH-501
					HDDR					
BATH RECORDER	SSQ-36	RO-308	SSQ-36	RO-308			SSQ-36	SSQ-36	SSQ-36	
HYPERBOLIC FIX UNIT	---	---	---	---			947957-103	948957-103	---	
TIME CODE GEN	---	---	TD-900/A	---	TD-900/AS	TD-900/A	TD-900/A	TD-900/A	---	TD-5062/AS
SEA NOISE METERS	---	---	---	---	ASQ-901	(1) ID-1872A	ID-1872A	ID-1872A	---	
GEN XMTR GP (CASS)	---	---		---			ASA-76	ASA-76	---	
NON ACOUSTIC										
RADAR	APS-80B	APS-134	APS-80B	APS-80B	APS-115B	APS-115B	APS-115B	APS-115B	APS-115	APS-116
SCAN CONVERTER	---	---	---	---	ASA-69	ASA-69	ASA-69	ASA-69		CV-5130/AP
ESM	ALD-2/ULA-2	ALD-2B	ALD-2	ALD-2B	ALQ-78	(1) ALQ-78	ALQ-78A	ALQ-78	ALD-2	ALR-502
		ULA-2	ULA-2	ULA-2				---	ULA-2	---
MAD/AUTO DETECT	ASQ-10	ASQ-10	ASQ-10	ASQ-10	ASQ-81	ASQ-81	ASQ-81	ASQ-81	ASQ-10	ASQ-502
MAD COMPENSATOR	CN-191/ASQ-8	CN-191/ASQ-8	CN-191/ASQ-8	CN-191/ASQ-8	ASA-65(V)5	ASA-65(V)5	ASA-65(V)5	ASA-65(V)5	ASA-65	OA-5154/ASQ
SAD	---	---	---	---	ASA-64A	ASA-64	ASA-64	ASA-64	---	---
SELECTOR CONTROL	---	---	---	---	ASA-71	ASA-71	ASA-71	ASA-71	---	OK-5022/A
IRDS	---	AAS-36	AAS-36	---	ASA-36	ASA-36	(1) AAS-36	AAS-36	---	OR-5008/AA
VIDEO RECORDER	---	V1000AB/NW		---	V1000AB/NW	V1000AB-NW	(1) OA-8962	OA-8962	---	673922-101
SEARCHLIGHT	AVQ-2C	AVQ-2C	AVQ-2C	AVQ-2C	---		---	---	AVQ-2C	---
AUX IRDS DISPLAY	---	---	---	---	---	OD-159/A	(1) OD-159/A	OD-159/A	---	---
SIDE-LOOKING RADAR	---	---	---	---	---		---	---	---	APD-10(1)
HARD COPY RECORDER										
ARMAMENT										
PHOTO MISC.										
MISSILES/CONTROL	---	ARW-77			ALG-19	ALG-19	(1) ALG-19	ALG-19		---
SURVEILLANCE CAM	KS-84			CANNON	CANNON		---	---	KA-74	KS-501A
SPEC PUR CAMERA	KB-16A	KB-10A		KB-104		(1) KB-18A	(1) KB-18A	(1) KB-18A	KB-18A	KL-501A
		KB-16A		KB-16A	KB-18A					
CRASH LOCATOR	---	URT-26	URT-26				URT-26	URT-26	---	URT-26
					URT-26	URT-26				
FLIGHT RECORDER	---	ASH-20	---	---	---		---	---	---	ASH-502

NOTES

(1) PROVISIONS ONLY

--- INDICATES DOES NOT APPLY TO CUSTOMERS AIRCRAFT

TABLE 3-B FMS P-3 AVIONICS CONFIGURATIONS

AVIONICS FUNCTION	SAF P-3A	RNZAF P-3B	RNOAF P-3B	RAAF P-3B	RAAF P-3C	RAAF P-3C	RNLN P-3C	JMSDF P-3C	IRAN P-3F	CAF CP-140
DPS & DISPLAYS										
DIGITAL COMPUTER	---	---	---	---	ASQ-114(V)4	ASQ-114(V)4	ASQ-114(V)4	ASQ-114(V)4	---	AYK-502
DATA PROCESSING		UDACS	ASN-124		AYA-8B	AYA-8B	AYA-8B	AYA-8B		
TACTICAL DISPLAY	ASA-16	UDACS	ASA-66B	ASA-16	ASA-70	ASA-70	ASA-70	ASA-70	ASA-16	OD-5006/A
TACT BEARING COMP	ASA-50	UDACS	---	ASA-50	---	---	---	---	ASA-50	---
NAVIGATION DISPLAY	PT-396	PT-396	PT-396	PT-396	ASA-70	ASA-70	ASA-70	ASA-70	PT-396	OD-5006/A
SENSOR DISPLAY	APA-125A		APA-125A	APA-125A	ASA-70	ASA-70	ASA-70	ASA-70	APA-125A	OD-5006/A
SENSOR DISPLAY	---	---	---	---	ASA-66	ASA-66	ASA-66	ASA-66	---	OD-5006/A
PILOT DISPLAY	GTP-4/ASA-13A	UDACS	ASA-66A	GTP-4/ASA-13A	ASA-66	ASA-66	ASA-66	ASA-66	GTP-4/ASA-13A	OD-5006/A
INTERVAL COMPUTER	TD-441A	UDACS	TD-441A	TD-441A	---	---	---	---	TD-414A	---
DMTS	---	DESIGNAT. UNK.	---	---	RD-319A	ASH-33	ASH-33	ASH-33	---	RD-5027/ASH
POSITION INDICATOR	ASA-13	ASA-13	ASA-13	ASA-13	---	---	---	---	ASA-13	---
COMMUNICATIONS										
UHF TRANSCEIVER	ARC-51A	ARC-51A	ARC-51A	ARC-51A	ARC-143B	ARC-143B	ARC-143B	ARC-143B	ARC-514	
UHF SECURE VOICE				KY-28		(1) KY–58		(1)		ARC-514
VHF	ARC-101	ARC-101	COM 2000	ARC-101	ARC-101	618M-3A	ARC-197	ARC-197	ARC-101	618M-3A (ARC-513)
HF TRANSCEIVER	ARC-94	ARC-94	ARC-94	ARC-94	ARC-161	ARC-161	ARC-161	ARC-161	618T-4	ARC-512
HF COUPLER	490T	490T	490T	490T	CU-2070	CU-2070	CU-2070	CU-2070	490T	CU-2041/AR
HF RECEIVER	ARR-41	ARR-41	(1) ARR-41	ARR-51	---	---	---	---		---
HF SECURE			KW-7		(1) KW-7	(1) KW-7	(1) KW-7	(1) KW-7		(1) KW-7
TELETYPE	TT-264/AG	TT-264/AG	TT-264/AG	TT-264/AG	AGC-6	AGC-6	AGC-6	AGC-6	TT-264	AGC-501
DATA CONVERTER	CV-1053	CV-1053	CV-1053	CV-1053	---	---	---	---	CV-1503	CV-5127
DATA LINK	---	---	---	---	ACQ-5A	ACQ-5A	ACQ-5A	ACQ-5A	---	(1) KG-40
SECURE DATA LINK						(1) KG-40				
INTERCOM	AIC-22(V)	AIC-22(V)	AIC-22(V)	AIC-22(V)	AIC-22(V)	AIC-22(V)	AIC-22(V)	AIC-22(V)	AIC-22(V)	AIC-503
IFF	APX-68	APX-72	APX-72	APX-72	APX-72	(1) APX-100	APX-72	APX-72	APX-72	APX-77
IFF INTERROGATOR	APX-7	APX-7	APX-7	APX-7	APX-76A	APX-76A	APX-76A	APX-76A	APX-76A	APX-502
IFF TEST SET			TS-1843A	TS-1843A	TS-1843A		TS-1843A	TS-1843A	TS-1843A	TS-1843B
IACS	---	---	---	---	---	---	---	---	---	---
SATCOM (OTCIXS)	---		---	---	---	---	---	---	---	(1)
EMERGENCY RADIO	PRT-5		PRT-5	CRT-3	PRT-5	PRT-5	PRT-5	PRT-5	PRT-5	PRT-5
SYNCHRO CONVERTER	---		---	---	CV-2461A/A	CV-2461A/A	CV-2461A/A	CV-2461A/A	---	
NAVIGATION										
INERTIAL	ASN-42	LTN-72	LTN-72	ASN-42	ASN-84	LTN-72	LTN-72	LTN-72	LTN-51	ASN-505
DOPPLER NAVIGATION	APN-153	APN-153	APN-153	APN-153	APN-187	APN-227	APN-227	APN-227	APN-153	APN-510
LORAN	APN-70	LTN-211		GNS-500A	---	---	---	---		---
OMEGA	---		ARN-131	---	ARN-99	ARN-99	(1) ARN-99	ARN-99	---	ARN-511
AIRMASS COMPUTER	ASA-47		ASN-124	ASA-47	---		---	---		---
TAS COMPUTER	A/A-24G-9	A/A-24G-9	A/A-24G-9	A/A-24G-9	A/A-24G-9	A/A-24G-9	A/A-24G-9	A/A-24G-9	A/A-24G-9	A/A-24G-9
VOR SET	ARN-87	ARN-87	ARN-87	ARN-87	ARN-87		ARN-140	ARN-140	ARN-87	ARN-508
TACAN	ARN-52	ARN-52	ARN-118	ARN-52	ARN-84	ARN-118	ARN-118	ARN-118	ARN-84	ARN-504
ADF	ARD-13	ARD-13	ARN-83	ARN-83	ARN-83	ARN-83	ARN-83	ARN-83	ARN-83	ARN-510
UHF DR/OTPI	ARA-25A	ARA-25A	ARA-25A	ARA-25A	ARA-50	ARA-50	ARA-50	ARA-50	ARA-50	
MARKER BEACON	ARN-32	ARN-32A	ARN-32	ARN-32	ARN-32		ARN-140	ARN-140	51Z4	ARN-508
FLIGHT DIR					AJN-15	AJN-15	AJN-15	AJN-15	AJN-15	AJN-501
AHRS	ASN-50	ASN-50	ASN-50	ASN-50						

Appendix F
P-3 Orion Chronology

1957 - Lockheed proposes development of L-188 Electra airliner as the basis of a new land-based ASW platform to meet the US Navy's specification #146.

1958 - (May) the Navy awards Lockheed a research and development contract to develop a prototype based on the L-188 Electra

- (August 19) the Lockheed Orion aerodynamic prototype (mockup), with dummy MAD boom and underbelly weapons bay fairing, makes maiden flight

- (September) the Navy inspects the aerodynamic prototype Orion

- (October) the Navy lets a contract to Lockheed for the manufacture of long-lead items such as forgings and complicated tools

1959 - (February) the Navy awards Lockheed an additional funds to develop a production prototype aircraft.

- (November 25) the second Orion prototype aircraft flies with a shorter fuselage and most of the planned avionics and sensor systems

1960 - (October) an initial production contract is awarded to Lockheed for seven production aircraft to be evaluated by the Navy

1961 - (April 15) the first P3V-1 Orion flies

1962 - (April 15) Navy BIS trials begin at NATC Patuxent River

- (July / August) P3V-1 Orions are delivered into naval service with VP-8

- (August) VP-44 receives P3V-1 Orions

- (September) the P3V-1 are re-designated "P-3A" Orions

- (October) P-3 from VP-8, VP-44 and those still under evaluation with Air Development Squadron One (VX-1) become involved in the Cuban Missile Crisis

1963 - (April 30) P-3A Orions deploy for the first time over seas, to Argentia (Newfoundland, Canada) with VP-44 and Atzugi (Japan) with VP-31

1964 - (March) P-3A Orions deploy to the Pacific Fleet and up to the Arctic

1965 - (February) the Orion goes to Vietnam, supporting Market Time operations throughout Southeast Asia

- (April) P-3B model Orion production begins

1966 - (February) the first P-3B Orions enter Navy service beginning with VP-26 and later VP-9 (in November) deployed to both the Pacific and Atlantic Fleets

- (August) 5 P-3B Orions are purchased by New Zealand, thus becoming the first international Orion operator

1967 - (June) NASA receives the P-3 Orion prototype aircraft and converts it into a NP-3A remote-sensing platform

1968 - (January) Australia becomes the second international Orion operator with the purchase of ten P-3B aircraft for its #11

Squadron

- (September 18) the YP-3C prototype of the Charlie Orion, equipped with the "A-NEW" System, flies for the first time

1969 - (April) Norway becomes the next Orion operator with the purchase of five P-3B Orions

- (June) the first production P-3C Orions enter service with VP-56 and later in the year with Fleet Replacement Squadron 30

- (June) the EP-3B Orions begin service with VQ-1

1970 - the P-3 Orion begins service with reserve patrol squadrons

- Lockheed finalizes a contract to convert four P-3A into "WP-3A" weather reconnaissance aircraft

- (July) the new P-3C Orion deploys to ICELAND for the first time with VP-49 and later to ALASKA with VP-47

1971 - (January 29) Weather Reconnaissance Squadron Four (VW-4) receives the first of four WP-3A Orions

- (August) NADC Warminster flies the P-3A / S-3A avionics testbed aircraft for the first time

- RP-3D Project Magnet is delivered to Oceanographic Development Squadron Eight (VXN-8)

1972 - (November 4) VXN-8 set's a class "C" / Group II world distance -closed circuit record, covering 5445.6 NM in sixteen hours (non-refueled)

1973 - (August) Spain takes delivery of three Deltic P-3A Orions

- (October) US Navy P-3 Orions fly last Market Time support flights of Vietnam Conflict

1974 - (April) VX-1 conducts first tests of P-3C "Update I" configured Orion

1975 - Iran places an order for six P-3C Orions, subsequently designated "P-3F" Orions

- (January) the P-3C Update I is incorporated in to Lockheed Orion production

- (April 30) VW-4 disestablished with WP-3A Orions subsequently being reworked into other Orion variants

- (June) NOAA / Department of Commerce receives their first (of two) "WP-3D" weather reconnaissance aircraft from Lockheed

- (July) the P-3C Update I enters fleet service with VP-19

1976 - (May) the P-3C Update I deploys to the Pacific for the first time with VP-19

- (July) Canada places an order for what subsequently becomes the "CP-140 Aurora"

1977 - Spain leases five P-3A from the US Navy

- (August) VX-1 and NATC's ASW Aircraft Test Directorate conduct the first flight tests of the P-3C "Update II" Orion

1978 - (February) Australia orders ten P-3C (Update II) Orions for their #10 Squadron

- (June) the P-3C Update II enters fleet service with VP-44

- (November) the last maritime patrol missions of the P-3A, with an active squadron, are conducted

- Japan selects the P-3C Orion as it's replacement maritime patrol aircraft, with Kawasaki Heavy Industries licensed to assemble the aircraft

1979 - (March 22) the CP-140 Aurora prototype conducts it's first flight

1980 - (May) the first CP-140 Aurora is delivered to Canada with the remaining seventeen aircraft turned over during the following ten months

- Norway receives two used P-3B Orions from the US Navy

1981 - (April) Japans JMSDF receives three P-3C Update II Orions from Lockheed

- (November) the first P-3C Update II.5 debuts with the US Navy

- (November) the first of thirteen P-3C Orions (actually Update II.5) arrives in the Netherlands

1982 - (March) Kawasaki flies the first of it's P-3C Orion "knock-down" aircraft

1983 - the US Customs Service converts four ex-Navy P-3A Orions into drug interdiction aircraft to combat the war on drugs

1984 - (May) the Update III production version of the P-3C Orion debuts on the production line and is introduced into Navy service later in the year

- (June 14) the Lockheed produced P-3 AEW&C prototype makes its maiden flight at Palmdale California

- Australia receives the second batch of ten P-3C Update II (actually II.5 configuration) replacing the original ten P-3B Orions with #11 Squadron

1985 - (December) the first "TP-3A" Orions are converted and delivered to FRS squadrons VP-30 and VP-31

1986 - (April) Lockheed is awarded a Navy contract to convert low hour NUD P-3C into "EP-3E" ARIES II SIGNIT aircraft

- (November) the APS-137 ISAR radar deploys for the first time on a P-3C Update III with VP-44

- Norway orders four new P-3C Update III Orions, replacing its older P-3 Bravos

1987 - (June 12) the US Customs Service places an order for Lockheed's P-3 AEW&C

- Lockheed Field Mod teams begin producing "retrofitted P-3C Update III" from older Charlie Orion aircraft

1988 - (April 8) Lockheed's P-3 AEW&C prototype, complete with all avionics, makes first flight

- (June 17) US Customs Service receives first P-3 AEW&C

- (October) Lockheed is awarded a contract to develop the next-generation ASW platform for the US Navy, to replace the Orion

1989 - (February) the P-3C Update III deploys for the first time, with VP-45 to Bermuda

- (March 31) VP-62 transitions to the P-3C Update III, the first reserve squadron to receive modern front line equipment

- (April) US Customs receives their second P-3 AEW&C Orion

- (June 30) Canada orders the last three Orion aircraft of the Lockheed California production line, subsequently designated "CP-140A Arcturus"

- (July 13) Lockheed announces that the P-3 Orion production line will end in 1991

- (August) Norway receives the first of four new P-3C Update III Orions

- (December) Lockheed reports technical problems with the development of the P-3 Orion replacement aircraft the "P-7A"

- the P-3B TACNAVMOD training syllabus at VP-31 ends

1990 - (April) Lockheed delivers the last Orion aircraft to the US Navy

- (March 22) VP-66 flies the last operational mission of a US Navy P-3A Orion

- (July 20) with technical problems and the vacuum created with the end of the Cold War, the US Navy cancels its contract with Lockheed for the P-7A

- (August) Thailand requests five ex-Navy P-3A Orions, to be subsequently designated P-3T Orions

- (December) South Korea orders eight new P-3C Update III Orions, establishing a new Orion production line in Lockheed's Georgia production facility due to the Company's restructuring and closing of its California facilities

- Pakistan receives the three P-3C Update II.75 Orions it ordered previously, but are subsequently frozen from delivery (Pakistan doesn't finally get the Orions until 1997)

1991 - US Navy P-3 Orions become the first aircraft on scene with the outbreak of the Gulf War with Iraq and take up an immediate active role in the conflict

- (February 10) the US Navy announces that four active patrol squadrons; VP-19, VP-44, VP-48 and VP-56 are to be disestablished

- (March) two US Navy P-3C Update III Orions collide in mid-air sixty miles off the San Diego coast, with 27 crewmen lost

- (May) Japan's Kawasaki HI production facility produces the first of several variant aircraft based on the JMSDF P-3C Orion; an EP-3 !

- (October) Norway receives the first of two "P-3N" Orions it had modified from two of its P-3 Bravos by NADEP Jacksonville

- Korean P-3C Update III Orions commences production at Lockheed's Marietta Georgia facility

- Lockheed proposes a new next-generation MPA platform to be based on the Korean P-3C Update III airframe and dubbed "ORION II"

1992 - Canada receives its three CP-140A Arcturus aircraft

- the US Navy announces additional patrol squadrons to be retired including VP-6, VP-22, VP-49 and VP-31 (other units ; VP-17, VP-23, VP-24, VP-50 and reserve units VP-60, VP-67, VP-90 and VP-91 would all be retired between 1993 and 1995)

1993 -

1994 - (June 28) the first Georgia built P-3C Orion rolls out of final assembly and flies for the first time

1995 - the first Korean P-3C Update III Orion is delivered to the Korean Navy

- (Feburary 1) NAWC Orion aircraft are re-designated NP-3D

1996 -

1997 - (August 1) VQ-11 is established at NAS Brunswick and assigned the two EP-3J C^3I counter-measures training Orions

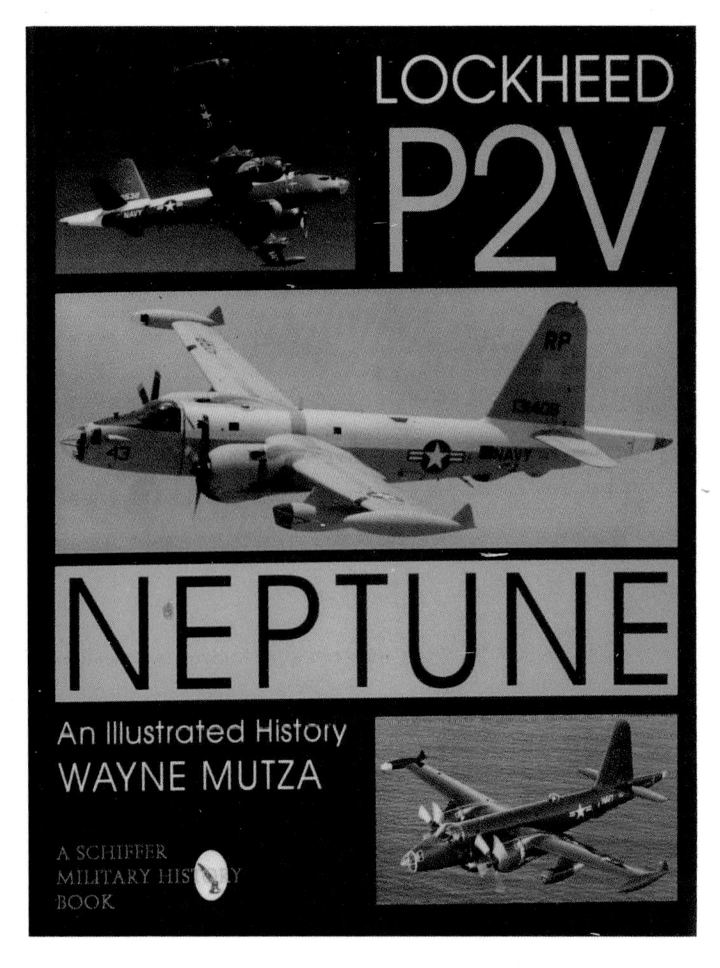

LOCKHEED P2V NEPTUNE: AN ILLUSTRATED HISTORY

Wayne Mutza

This long overdue account provides an extraordinary amount of insight into the Neptune's lengthy history, beginning with its inception during World War II to the present day survivors. More than 1,000 examples were built, many of which thrive today as fire bombers and warbirds. Presented here for the first time are the many fascinating details describing Neptune service with non-U.S. air arms and obscure operations. Clearly evident is the in-depth research that makes this extensive volume accurate, detailed and readable.

Size: 8 1/2" x 11"

over 400 b/w & color photographs, approx. 176 pages, hard cover

ISBN: 0-7643-0151-9 $49.95

WARBIRDS OF THE SEA: A HISTORY OF AIRCRAFT CARRIERS & CARRIER-BASED AIRCRAFT

Walter A. Musciano.

Covers the history and combat career of aircraft carriers and shipboard aircraft from their conception into the future.

Size: 8 1/2" x 11"

over 800 b/w photos, drawings, maps

592 pages, hard cover

ISBN: 0-88740-583-5 $49.95